BATTLEGROUND
ENVIRONMENT

BATTLEGROUND

ENVIRONMENT

VOLUME 2 (I–Z)

Robert William Collin

GREENWOOD PRESS
Westport, Connecticut • London

Library of Congress Cataloging-in-Publication Data

Collin, Robert W., 1957–
   Battleground : environment / by Robert William Collin.
      p. cm.
   Includes bibliographical references and index.
   ISBN-13: 978–0–313–33865–6 (set : alk. paper)
          978–0–313–33866–3 (v. 1 : alk. paper)
          978–0–313–33867–0 (v. 2 : alk. paper)
1. Environmental sciences.   2. Environmental policy.   3. Environmental degradation.
4. Human beings—Effect of environment on.   I. Title.   II. Title: Environment.
   GE105.C65   2008
   363.7—dc22   2008002114

British Library Cataloguing in Publication Data is available.

Library of Congress Catalog Card Number: 2008002114
   ISBN: 978–0–313–33865–6 (set)
   ISBN: 978–0–313–33866–3 (vol. 1)
   ISBN: 978–0–313–33867–0 (vol. 2)

First published in 2008

Greenwood Press, 88 Post Road West, Westport, CT 06881
An imprint of Greenwood Publishing Group, Inc.
www.greenwood.com

Printed in the United States of America

The paper used in this book complies with the
Permanent Paper Standard issued by the National
Information Standards Organization (Z39.48–1984).

10 9 8 7 6 5 4 3 2 1

These books are dedicated to the spirit and vision
of the late Damu Smith, founder of the National Black Environmental
Justice Network, and former Greenpeace community organizer
in Louisiana and Texas. He recognized and taught that the
environmental context of Truth can bring us together to face the
environmental controversies in our midst.

Battleground: Environment is also dedicated to my two faithful
canine research companions, Ambar and Max.

CONTENTS

GUIDE TO RELATED TOPICS

AGRICULTURE AND ENVIRONMENTAL CONTROVERSIES
Genetically Modified Food
Hemp
Industrial Agricultural Practices and the Environment
Industrial Feeding Operations for Animals
Organic Farming

ANIMALS AND ENVIRONMENTAL CONTROVERSIES
Animals Used for Testing and Research
Arctic Wildlife Refuge and Oil Drilling
Cultural vs. Animal Rights: The Makah Tribe and Whaling
Endangered Species
Preservation: Predator Management in Oregon
Wild Animal Reintroduction

CHILDREN AND ENVIRONMENTAL CONTROVERSIES
Cancer from Electromagnetic Radiation
Cell Phones and Electromagnetic Radiation
Childhood Asthma and the Environment
Children and Cancer

CITIZEN ENVIRONMENTAL CONTROVERSIES
Citizen Monitoring of Environmental Decisions
Collaboration in Environmental Decision Making

HUMAN ENVIRONMENTS
Brownfields Development
Environmental Regulation and Housing Affordability
Environmental Vulnerability of Urban Areas
Poverty and Environment in the United States
Transportation and the Environment

IMPACT ASSESSMENT
Environmental Impact Statements: International
Environmental Impact Statements: Tribal
Environmental Impact Statements: United States

INDUSTRY PRACTICES AND ENVIRONMENTAL CONTROVERSIES
Permitting Industrial Emissions: Air
Permitting Industrial Emissions: Water
Persistent Organic Pollutants
Pesticides

LAND USE AND ENVIRONMENTAL CONTROVERSIES
Big-Box Retail Development
Evacuation Planning for Natural Disasters
Federal Environmental Land Use
Good Neighbor Agreements
Land-Use Planning in the United States
Ski Resort Development and Expansion
Sprawl
State Environmental Land Use
"Takings" of Private Property under the U.S. Constitution
Watershed Protection and Soil Conservation

NATURAL DISASTERS
Avalanches
Drought
Floods
Hurricanes
Ice
Landslides and Mudslides
Mountain Rescues
Tsunami Preparation

I

ICE

Controversies around ice revolve around the accurate prediction of ice storms, responsibility for cleanup and power restoration, and responsibility for keeping ports and harbors clear of ice. With the coming climatic changes, natural disasters and controversies due to ice could increase.

WHAT IS ICE?

Ice is frozen water. Its environmental impact is intrinsic to ecosystems and damaging to human systems. Ice can reach anywhere water can reach. When water becomes ice, it generally expands with force. Ten percent of the land surface is currently covered with ice, although global warming is shrinking the ice caps. In human habitation, ice can have a very destructive impact on the built environment—roads, bridges, buildings, pipes, and sewers. Water and ice are also very heavy. When ice coats objects they become very heavy. Tree limbs break and fall on buildings and power lines. Power lines themselves become heavy and fall, often leaving live electrical currents on the ground. Roads, rail lines, and airports shut down. Without electricity many heating systems as well as food and medical refrigeration systems fail, often at a time they are most needed. The eastern United States and Canada can experience freezing rain any time between late October and early May; however, trends in global warming and climate change may extend the season and geography for ice storms.

Ice storms in North America develop along a line stretching from northern Texas and western Mississippi through the Midwest and mid-Atlantic states and throughout the northeastern states. Most of eastern Canada is subject to ice storms. They usually require a substantial, slow-moving low-pressure system with a large temperature difference between the warm Gulf air and cold Arctic air. Ice storms can last as long as these two air systems persist with temperature differentials and moisture. If they clash around the Great Lakes they can accumulate ice for a long time.

The temperature of freezing will vary with the size of the water droplet and the concentration of any impurities in it. If the water has formed around pollutants and particulate matter, then it is very difficult to assess the actual freezing point, which many current weather models rely on for forecasting these events. This is a point of scientific controversy. Very small droplets of pure water may not freeze spontaneously until the temperature has fallen to around −40°C (−40°F).

Many scientists and most communities want more monitoring of weather for use in many cases, ice disasters being one of them. Governmental agencies have typically been slow to respond with adequate or accurate monitoring that would be necessary for the level of certainty that current scientific models of proof require for causality. Ice storms are fickle and very dynamic.

FREEZING RAIN IMPACTS

Even small amounts of freezing rain can increase car accidents and impair overall mobility. Ice can impair the ability of animals to get food and shelter. A thickly iced field will provide little escape for deer, elk, caribou, sheep, and cattle from predators and also little food for herbivores because of the ice coatings. Tree branches and trunks collect ice in vast quantities, called *ice loading*. Conifers are resistant to ice loading because of their flexibility, tapered shape, and lack of trunk branching. A 50-foot conifer tree can accumulate 99,000 pounds of ice during a severe storm. Deciduous trees cannot hold as much ice and break more frequently. Some trees are more fragile than others. Indigenous trees usually have a stronger resistance to regional weather patterns, including ice storms. Generally, the more large branches a tree has, the more likely it is to break from ice loading. Ice also seals off food for winter animals such as deer, elk, and rabbits.

POTENTIAL FOR FUTURE CONTROVERSY

Given global warming impacts, a strong reliance on modern technology, and inadequate monitoring of environmental conditions, it is likely that natural disasters and more controversies involving ice will continue. The battleground for this controversy is first subsumed in the larger controversies of global warming and climate change. Another battleground for this controversy is the use and production of electrical energy. Is the technology failing us, or is nature creating the disaster? Environmentalists and large business interests are lined up in opposing positions in both these battlegrounds. Communities and governments have a range of positions from none at all to complete engagement with environmental quality and/or economic development. Ice storms may be one of the first indicators of climate change and may change some of the positions of communities and their governments. That could radically shift the battleground, and controversy in courts and legislatures would ensue.

See also Climate Change; Floods; Global Warming

Web Resources

Ice Storms: Environmental Impacts. Available at kyclim.wku.edu/BRADD/icestorms/envi ron.html. Accessed January 21, 2008.

Statistics Canada. The Ice Storm 1998: Maps and Facts Activity. Available at www.statcan. ca/english/kits/storm.htm. Accessed January 21, 2008.

U.S. Environmental Protection Agency. Natural Disasters and Weather Emergencies: Snow and Ice. Available at epa.gov/naturalevents/snow-ice.html. Accessed January 21, 2008.

Further Reading: Doheny-Farina, Stephen. 2001. *The Grid and the Village: Losing Electricity, Finding Community, Surviving Disaster.* New Haven, CT: Yale University Press; Kaplan, Laura G. 1996. *Emergency and Disaster Planning Manual.* New York: McGraw-Hill Professional; MacGuire, Bill, Ian M. Mason, and Christopher R. J. Kilburn. 2002. *Natural Hazards and Environmental Change.* New York: Oxford University Press; Stoltman, Joseph P., John Lidstone, and Lisa M. Dechano, eds. 2005. *International Perspectives on Natural Disasters: Occurrence, Mitigation, and Consequences.* New York: Springer.

INCINERATION AND RESOURCE RECOVERY

Burning waste for energy purposes is controversial because it may increase the toxicity of the emissions in the form of ash. This is waste that would otherwise go to full U.S. landfills. Many incineration corporations claimed they were making energy to get status as a utility. At that time utilities were exempt from the right-to-know laws.

INCINERATION: MORE THAN JUST BURNING THE TRASH

Burning the trash is an old custom in many rural areas in the United States. Even today, the EPA estimates that private residential trash burning is a major source of pollution in the upper Midwest. Historically, burning trash and waste was an improvement over leaving it around or just placing it in heaps. Burning the trash reduced its volume and risk. Waste can be a vector for many public health risks, and it can attract vermin. Rats, mice, and other rodents can become vectors for diseases such as bubonic plague and spread the risk of deadly disease deep into human populations. Waste can also be fuel and shelter. U.S. pioneers in the Midwest used buffalo chips (waste) to build sod houses. Dried, this waste could be burned for fuel and heat. This type of waste was the main type of waste, as opposed to today's chemically enriched, multisubstance, and potentially toxic waste stream.

An estimated 14–16 percent of the U.S. waste stream is incinerated. It could be more because many industrial and military wastes are incinerated. Generally, waste is delivered to the incinerator. The movement of waste itself is peppered with battlegrounds. The waste may have come from long distances, from cities and waste transfer stations. As U.S. environmental consciousness has increased, more people are interested in where their waste goes and what environmental impact it has. Waste transfer stations can become a local land-use battleground and sometimes an environmental justice battleground. When landfills become full, waste must wait in waste transfer stations. The waste can be hazardous. Many communities fear that the transfer station will become a permanent waste site, as some have done. The waste is delivered by truck, ship, and railroad. Sometimes there are spills of hazardous wastes with severe environmental and community repercussions. Environmental justice communities may have an overconcentration of poorly regulated waste transfer sites. These sites are where many spills occur. When energy is to be used from the incineration, the waste is taken to an energy recovery facility where it is burned in combustion chambers or boilers. High combustion temperatures can help most of the waste burn thoroughly. This is one goal of incineration, less ash for disposal. Environmentalists are very concerned with air emissions from incinerators. They have large emissions that can accumulate quickly around the site as heavier particulate matter such as metals falls to the ground. Metals are difficult to burn completely. Much of what the incinerators put out reflects what is put into them. Waste stream control is difficult at best, although improvements in recycling have benefited other waste streams. Older, pre-1970 incinerators burned everything they could fit into them. Older buildings painted with lead paint are still burned as hazardous wastes. The lead does not burn and drops down as particulate matter, creating a risk of toxicity, as it did in Flint, Michigan. Combustible, explosive, illegal, and other dangerous materials inhabit the waste stream. Evidence of a crime or environmental impact that is burned is usually less recognizable than when placed in a landfill. Medical wastes can include all sorts of waste, including trace amounts of radioactive material. Many older municipal incinerators in the southeastern United States were placed in African American communities, exposing generations to decades of heavy metals such as lead, mercury, and cadmium. Battlegrounds such as this surround the movement of waste generally, and incineration of it specifically. Industry claims that modern air pollution control devices include electrostatic precipitators, dry and wet scrubbers, and/or fabric; and that these get out everything dangerous. Cumulative emissions are not measured, although their human and environmental impacts can be large.

CURRENT WASTE MANAGEMENT IN THE UNITED STATES

To say that waste is managed is an overstatement. The main characteristic of the U.S. waste trends is the massive increase in volume. Enormous progress in regulating the waste stream is a primary characteristic of the U.S. waste management approach. Federal government financing of the expensive infrastructure, sometimes inclusive of modern pollution-control and abatement technologies, made it possible for local governments to treat solid wastes. The waste stream of the 2000s is very different than 50 years ago. It contains inorganic materials that may create risks for people and for the environment. This has increased the overall controversy of siting and permit renewals for incinerator facilities. It has also moved the battleground into the courts. Recently, incinerators in the African American community of urban north Florida from the 1920s until the 1970s provided the basis for a successful $76 million tort settlement, for wrongful deaths and other environmental impacts. The plaintiffs are quick to point out that money does not replace lost lives of loved ones. These bitter victories only increase the rancor of this controversy.

THE BENEFITS OF RESOURCE RECOVERY THROUGH INCINERATION

Industry claims that by burning the solid wastes into ash, incineration reduces the volume of waste entering the landfill by approximately 90 percent.

ENERGY VALUE OF PLASTICS, MUNICIPAL SOLID WASTES, AND NATURAL RESOURCES

Plastic Material Energy Value

When material is incinerated it gives off different levels of energy useful in resource recovery. Emissions from burning these materials, their amount, toxicity, and the remaining ash are not considered.

Polyethylene terephthalate	9,000–9,700
Polyethylene	19,900
Polyvinyl chloride	7,500–9,000
Polypropylene	18,500–19,500
Polystyrene	17,800

Municipal Solid Waste Material Energy Value

Newspaper	8,000
Textiles	6,900
Wood	6,700
Yard wastes	3,000
Food wastes	2,600
Average for municipal solid waste	4,500

Natural Resources Energy Value

Fuel oil	20,900
Wyoming coal	9,600

Plastics are pervasive in U.S. society. They are used all over the world in agriculture in large amounts. Incineration of wastes such as plastic to create electrical energy could be sustainable depending on the environmental impacts of the ash and air emissions.

Recovering some of the energy from burning waste can produce electricity. This can help offset any potential cost of environmental mitigation. Resource recovery by incineration of wastes is considered so efficient that old landfills are being opened up and that waste then incinerated for its energy potential.

RESOURCE RECOVERY OF PLASTICS

Plastics have a higher energy value and heat content than most municipal solid waste materials. While making up 7 percent of the waste stream by weight and 20 percent by volume, plastics provide incinerators with 25 percent of the recoverable energy from municipal solid wastes. A pound of polyethylene supplies 19,000 btu, but corrugated paper packaging provides only 7,000 btu. Incinerating plastics produces more energy.

Another major battleground around incinerating plastics is the environmental impacts of the emissions. Many plastics contain heavy metals such as lead and cadmium, which might increase the toxicity of the incinerator ashes. The metal content could cause the ashes to be hazardous wastes. If they are emitted into the air, then they may fall as particulate matter on nearby land and waterways. Currently, incinerator ashes are not categorized as hazardous wastes and can usually be disposed of in landfills. In some communities this can be a large battleground. When this happens, this ash can back up in waste transfer stations. The waste transfer stations are not designed or sited as a terminal place for waste of any kind, much less hazardous wastes. A small incineration plant processes approximately 300 tons of waste each day while a large plant can process about 3,000 tons.

Incinerators can produce enough energy to run an industrial facility or a small community, depending on volume and kind of waste stream.

WASTE-TO-ENERGY FACILITIES AND THEIR OPERATIONS

The cost of building a waste-to-energy incinerator is very high. The volume of truck traffic will increase. Depending on the facility's permit to operate, it may not be approved to accept all wastes. Trucks delivering wastes are required to display signs that indicate the hazards of their waste loads. Economically, resource-recovery incinerators need a reliable, steady stream of high-energy waste or else processing the waste material will cost more energy than it produces.

POTENTIAL FOR FUTURE CONTROVERSY

Incineration is a waste treatment method sanctioned by the government because the research is inadequate to prove it unsafe. This lack of science allows potentially dangerous emissions to enter the environment and community. The effect of the quantity and types of pesticides in the incineration process has yet to be determined, and may present a presently unaccounted exposure vector to nearby populations and environments. Many communities feel that the burden should be on industry to prove an emission is safe before it is allowed. Currently the burden is on those harmed to prove it is unsafe to a scientific level of certainty,

INCINERATION AS A METHOD FOR RESOURCE RECOVERY FROM INEDIBLE BIOMASS IN A CONTROLLED ECOLOGICAL LIFE SUPPORT SYSTEM

In research published by the NASA Ames Research Center Advanced Life Support Division, Regenerative Systems Branch, Moffett Field, California, waste recovery is serious business.

Resource recovery from waste streams in a space habitat is essential to minimize the resupply burden and achieve self-sufficiency. In a controlled ecological life support system (CELSS) human wastes and inedible biomass will represent significant sources of secondary raw materials necessary for support of crop plant production (carbon, water, and inorganic plant nutrients). Incineration, pyrolysis, and water extraction have been investigated as candidate processes for recovery of these important resources from inedible biomass in a CELSS. During incineration carbon dioxide is produced by oxidation of the organic components, and this product can be directly utilized by plants. Water is concomitantly produced, requiring only a phase change for recovery. Recovery of inorganics is more difficult, requiring solubilization of the incinerator ash. The process of incineration followed by water solubilization of ash resulted in the loss of 35 percent of the inorganics originally present in the biomass. Losses were attributed to volatilization (8%) and non–water-soluble ash (27%). All of the ash remaining following incineration could be solubilized with acid, with losses resulting from volatilization only. The recovery for individual elements varied. Elemental retention in the ash ranged from 100 percent of that present in the biomass for Ca, P, Mg, Na, and Si to 10 percent for Zn. The greatest water solubility was observed for potassium, with recovery of approximately 77 percent of that present in the straw. Potassium represented 80 percent of the inorganic constituents in the wheat straw and, because of slightly greater solubility, made up 86 percent of the water-soluble ash. Following incineration of inedible biomass from wheat, 65 percent of the inorganics originally present in the straw were recovered by water solubilization and 92 percent by acid solubilization. Recovery of resources is more complex for pyrolysis and water extraction. Recovery of carbon, a resource of greater mass than the inorganic component of biomass, is more difficult following pyrolysis and water extraction of biomass. In both cases, additional processors would be required to provide products equivalent to those resulting from incineration alone. The carbon, water, and inorganic resources of inedible biomass are effectively separated and output in usable forms through incineration.

which is quite high. This controversy reopens a more basic question of whether science should control policy. If the scientific methods are underfunded, slow, and not accessible, then environmental policy stagnates. However, political pressure often forces the government to develop new controversial policies without waiting for the science to catch up. Incineration as an environmental practice by itself may be limited as environmental policy expands to include concepts of sustainability. However, because it uses the energy from waste and diverts waste from landfills, resource recovery from incineration may see new applications. The larger question is what to do with all the waste, how to stop the generation of waste, and how to clean up the waste already here. Controversies will follow each of these questions along the way.

See also Cumulative Emissions, Impacts, and Risks; Permitting Industrial Emissions: Air; Permitting Industrial Emissions: Water; Sustainability; Toxics Release Inventory

Web Resources

Biocrawler.com. Waste Incineration. Available at www.biocrawler.com/encyclopedia/Incineration. Accessed January 21, 2008.

Eco. Waste. Available at www.ecozine.co.uk/Waste.htm. Accessed January 21, 2008.

Further Reading: Dhir, Ravindra, Thomas D. Dyer, and Kevin A. Paine, eds. 2000. *Sustainable Construction: Use of Incinerator Ash.* London: Thomas Telford Publishing; Hamerton, Azapagic, Emsley, Adisa Azapagic, and Alan Emsley, eds. 2003. *Polymers: The Environment and Sustainable Development.* Hoboken, NJ: John Wiley and Sons; Holly, Hattemer-Frey, Janos Szollosi, Lajos Tron, and Curtis C. Travis, eds. 1991. *Health Effects of Municipal Waste Incineration.* Boca Raton, FL: CRC Press; Niessen, Walter R. 2002. *Combustion and Incineration Processes: Applications in Environmental Engineering.* New York: Marcel Dekker; Smith, D. Clayton, Rex H. Warland, and Edward J. Walsh. 1997. *Don't Burn It Here: Grassroots Challenges to Trash Incinerators.* University Park, PA: Pennsylvania State University Press.

INDIGENOUS PEOPLE AND THE ENVIRONMENT

Indigenous people have a long-term and ancient relationship with the environment. Nonindigenous relationships with the environment can conflict with these traditions, and cause controversy.

Indigenous people have many of the same controversies around the environment as do other communities. Throughout history, indigenous peoples have maintained a strong connection with the environment, a connection that is integral to the survival of their physical, social, economic, cultural, and spiritual ways of life. Because of this connection, they will likely be disrupted by climate change impacts more severely than many other citizens.

In the United States, many indigenous people were forced to move to reservations as a condition of tribal recognition by the U.S. government. Those that did not are called bands. Most bands were hunted and destroyed by European settlers. Tribes in the United States have some sovereignty over their lands and therefore control of the environment. Some tribes pursue status as states to issue their own water regulations. Most often, tribes were driven and coerced into places no one else wanted. There are about 500 tribes in the United States, of which only about 10 percent have any casino revenue. In terms of environmental controversies, they include the following.

- Toxic contaminants, agricultural pesticides, and other industrial chemicals that disproportionately impact indigenous peoples, especially subsistence and livestock cultures.
- Inadequate governmental environmental and health standards and regulations.
- Cleanup of contaminated lands from mining, military, and other industry activities.

- Toxic incinerators and landfills on and near indigenous lands.
- Inadequate solid and hazardous waste and wastewater management capacity of indigenous communities and tribes.
- Unsustainable mining and oil development on and near indigenous lands.
- National energy policies at the expense of the rights of indigenous peoples.
- Climate change and global warming.
- Coal mining and coal-fired power plants resulting in mercury contamination, water depletion, destruction of sacred sites, and environmental degradation.
- Uranium mining developments and struggles to obtain victim compensation for indigenous uranium miners, millers, processors, and those downwind of past nuclear testing experiments.
- Nuclear waste dumping in indigenous lands.
- Deforestation.
- Water rights, water quantity, and privatization of water.
- Economic globalization putting stress on indigenous peoples and local ecosystems.
- Border justice, trade agreements, and transboundary waste and contamination along the U.S./Mexico/Canada borders and other indigenous lands worldwide.
- Failure of the U.S. government to fulfill its mandated responsibility to provide funding to tribes and Alaskan villages to develop and implement environmental protection infrastructures.
- Backlash from U.S. state governments giving in to the lobbying pressure from industry and corporations against the right of tribes to implement their own water and air quality standards.
- Protection of sacred, historically and culturally significant areas.
- Biological diversity and endangered species.
- Genetically modified organisms impacting the environment, traditional plants and seeds and intellectual rights of indigenous peoples—biocolonialism.
- Economic blackmail and lack of sustainable economic and community development resources.
- Just transition of workers and communities impacted by industry on and near indigenous lands. This can refer to training and educational programs.
- Urban sprawl and growth on and near indigenous lands.
- Failure of colonial governments and their programs to adequately consult with or address environmental protection, natural resource conservation, environmental health, and sacred/historical site issues affecting traditional indigenous lands and indigenous peoples.
- Decolonization and symptoms of internalized oppression/racism/tribalism. This can refer to self-destructive behaviors and loss of cultural identity.

INDIGENOUS PEOPLE AND ENGAGEMENT WITH ENVIRONMENTAL CONTROVERSY

Indigenous people all over the world engage in environmental struggles. The environmental battleground is diverse. Here is a short, partial list of some indigenous organizations and the environmental controversies they focus on, as reported by the Indigenous Environmental Network.

The Midwest Treaty Network is an alliance of Indian and non-Indian community groups that support the sovereign rights of Native American nations. While founded in the context of the Chippewa (Ojibwe) treaty struggle, it is concerned generally with defending and strengthening native cultures and nationhood, protecting Mother Earth, and fighting racism and other forms of domination throughout that region. The network has taken a stand against economic and political pressure on indigenous nations to give up their rights.

The mission of the Haudenosaunee Environmental Task Force (HETF) is to assist Haudenosaunee nations in their efforts to conserve, preserve, protect, and restore their environmental, natural, and cultural resources; to promote the health and survival of the sacred web of life for future generations; to support other indigenous nations who are working on environmental issues; and to fulfill responsibilities to the natural world as the creator instructed without jeopardizing peace, sovereignty, or treaty obligations. The leaders of the Haudenosaunee have always considered three principles when making decisions: will a decision threaten peace, the natural world, or future generations?

The Assembly of First Nations (AFN) is the national representative/lobby organization of First Nations in Canada. There are over 630 First Nations communities in the country. The AFN Secretariat is designed to present the views of the various First Nations through their leaders.

The Indigenous Environmental Network (IEN) is a grassroots alliance of indigenous peoples whose mission is to protect the sacredness of Mother Earth from contamination and exploitation by strengthening, maintaining, and respecting the traditional teachings and the natural laws, as well as by building sustainable communities. The IEN is not simply a combination of the Native American movement with environmental activism. IEN has popularized a new perspective on native sovereignty that includes appropriate technology and the defense of natural resources. IEN's perspective on environmentalism includes "supporting the survival of endangered cultures and putting the protection of nature in a larger social, cultural and economic context."

The goals of the National Environmental Coalition of Native Americans (NECONA) are:

- To educate Indians and non-Indians about the health dangers of radioactivity and the transportation of nuclear waste on America's rails and roads.
- To network with Indian and non-Indian environmentalists to develop grassroots counter-movement to the well-funded efforts of the nuclear industry.
- To declare tribal nuclear-free zones across the nation.

The Indigenous Peoples Council on Biocolonialism (IPCB) is organized to assist indigenous peoples in the protection of their genetic resources, indigenous knowledge, and cultural and human rights from the negative effects of biotechnology.

The Coalition for Amazonian Peoples and Their Environment is an initiative born out of the alliance between indigenous and traditional peoples of the Amazon and groups and individuals who share their concerns for the future of the Amazon and its peoples. The 80 nongovernmental organizations from the north and the south that are active in the coalition believe that the future of the Amazon depends on its indigenous and traditional peoples and the state of their environment.

POTENTIAL FOR FUTURE CONTROVERSY

All groups of indigenous people have their own unique environmental controversies. However, the environment knows no human political boundary, and ecosystem pollution can affect everyone in contact with the land, air, and water. As long as global warming, climate change, and cumulative ecosystem effects continue, it is likely that environmental controversies will engage indigenous people.

As the U.S. population rises and uses more natural resources, such as water, conflict with indigenous people may increase as they seek to enforce their rights under various treaties and agreements. Tribes can get status as states and develop their own water-quality standards. This can affect the number of industries getting permits to emit chemicals into the water, even if they are off reservation. The battleground for this type of controversy now is the southwestern United States and other areas with scarce water.

See also Climate Change; Cultural vs. Animal Rights; Cumulative Emissions, Impacts, and Risks; Ecosystem Risk Assessment; Environmental Impact Statements: Tribal; Genetically Modified Food; Sacred Sites

Web Resources

Amazon Watch. 2003. Project Profile: The OCP Pipeline. Available at www.rainforestinfo. org.au/ocp/background.htm. Accessed January 21, 2008.

Indian Country. 2004. Nature Conservancy Efforts Disregard Indigenous Peoples. Available at www.indiancountry.com/content.cfm?id=1096409914. Accessed January 21, 2008.

Indigenous Environmental Network. Available at www.ienearth.org/. Accessed January 21, 2008.

Further Reading: Fixico, Donald Lee. 2003. *The American Indian Mind in a Linear World: American Indian Studies and Traditional Knowledge.* New York: Routledge; Gedicks, Al. 1993. *The New Resource Wars: Native and Environmental Struggles against Multinational Corporations.* Boston: South End Press; Harkin, Michael Eugene, and David Rich Lewis. 2007. *Native Americans and the Environment: Perspectives on the Ecological Indian.* Lincoln: University of Nebraska Press; Johnson, Troy R. 1999. *Contemporary Native American Political Issues.* MD: Rowman Altamira; Krech, Shepard, III. 2000. *The Ecological Indian: Myth and History.* New York: W. W. Norton and Company; Selin, Helaine. 2003. *Nature across Cultures: Views of Nature and the Environment in Non-Western Cultures.* New York: Springer.

INDUSTRIAL AGRICULTURAL PRACTICES AND THE ENVIRONMENT

Industrial agricultural practices can have large impacts on the environment and entire ecosystem. Environmentalists, small family farmers, downstream communities, and environmental justice communities object to these impacts.

Historically, humans as hunter-gatherers would hunt in an area and move on. Gradually, a human community would farm a given location. When the game was gone and the soil infertile, the community would move to another location. Human population was so small then, and the environmental impacts of the technology so low, that this allowed the used-up region to regenerate. This method was sustainable only so long as there were new places to move to and environmental impacts remained within the period of time necessary for regeneration of natural systems. This method was used by European colonial powers all around the planet. In the United States this method was used by European farmers who moved to the New World. Many European settlers to North America felt a manifest destiny to colonize the continent from coast to coast.

E. COLI IN FOOD AS A RESULT OF INDUSTRIAL AGRICULTURAL PRACTICES?

E. coli is everywhere in the environment. Some strains can be deadly. At the Lane County Fair in Eugene, Oregon, about 140 children contracted a severe strain of E. coli. All the children survived, but some were hospitalized for an extended period. All of them now have that strain within them. The cause of this contact was the failure of children to wash their hands after petting the goats and sheep.

Another recent concern is that some strains of E. coli could be the result of agricultural industrial practices. It has been found in the soil and vegetables from Californian's central valley. As these farms are not rigorously regulated for these risks, there is much controversy about who is to blame for them.

INDUSTRIALIZED AGRICULTURE

Industrialized agriculture can have several meanings. It can mean substituting machines for people in the food production process, increasing the scale of production beyond the regenerative capacity of the land, and using chemicals instead of natural organic materials. In the case of chemicals, when farmers discovered that certain chemicals can replace the older way of fertilizing, they realized they could save time. The old process of fertilizing was called *manuring*. It took a large amount of time. Farmers in search of higher productivity industrialized in order to compete on world markets. They use more machines, increase the scale of production, and rely on technology for time-saving efficiency in food production and preparation. Unfortunately, these machines can emit environmentally damaging pollutants, the land can give out, and the chemicals can create public health risks. Each category of industrialization of agriculture is a battleground within this controversy and is of concern to environmentalists and others because of its potential environmental impacts and human health risks.

The contours of this controversy are shaped by rapidly developing technology that thrives on large-scale applications and a growing scientific consensus and mobilization of community concern about health risks.

Mainstream agriculture faces enormous controversies and dynamic changes driven in part by environmental issues. The battleground for this controversy is affected by the following changing conditions.

- Climate change will have a major impact on agricultural practices in many areas.
- Agriculture will have to find alternative energy sources to sustain productivity because of current high reliance on nonrenewable energy.
- Environmental waste sinks are increasing in size. The hypoxic zone in the Gulf of Mexico increased to 8,200 square miles in 2002. Most scientists attribute this to runoff from agricultural activities all along the Mississippi River watershed. The same is true for most coastal outlets in industrialized nations.

SUSTAINABLE AGRICULTURE

Sustainable agriculture is defined as the ability to maintain productivity of the land. In general terms, sustainable agriculture includes the following principles.

- Ecologically restorative
- Socially resilient
- Economically viable

This shifts the emphasis from managing resources to managing ourselves. Agricultural corporations and associated trade groups view this as increased governmental intrusion and do not embrace these ideals in their entirety.

The small farmer is a part of the settlement of the United States. Many laws were written to protect the small family farmer. Currently, many of these laws are used by agribusiness. Some environmentalists think they do so to hide environmental impacts. It has been very difficult to get right-to-know legislation passed in agricultural areas in agricultural states. Agribusiness resists the increased reporting requirements because of the added cost and decreased profitability, especially in regard to pesticides.

ANATOMY OF CENTERS FOR DISEASE CONTROL (CDC) RESPONSE TO *E. COLI*

On Friday, September 8, 2006, Centers for Disease Control (CDC) officials were alerted by epidemiologists in Wisconsin of a small cluster of *E. coli* serotype O157:H7 infections of unknown source. Wisconsin also posted the DNA fingerprint pattern of the cluster to PulseNet, thus alerting the entire network. Separately, the state health department of Oregon also noted a very small cluster of infections that day and began interviewing the cases. On September 13, both Wisconsin and Oregon reported to CDC that initial interviews suggested that eating fresh spinach was commonly reported by cases in both clusters of *E. coli* serotype O157:H7 infections in those states. PulseNet showed that the patterns in the two clusters were identical, and other states reported cases with the same PulseNet pattern among ill persons who also had eaten fresh spinach. CDC notified the Food and Drug Administration (FDA) about the Wisconsin and Oregon cases and the possible link with bagged fresh spinach. The CDC and FDA convened a conference call on September 14 to discuss the outbreak with the states.

Quick sharing of information among the states, CDC, and FDA led to the FDA warning the public on September 14, 2006, not to eat fresh bagged spinach. On September 15, the number of reported cases approached 100. Cases were identified by PulseNet and interviewed in detail by members of OutbreakNet. Leftover spinach was cultured at the CDC, FDA, and in state public health laboratories. The epidemiologic investigation indicated that the outbreak was from a single plant on a single day during a single shift.

Coordination with the FDA was important for investigating this outbreak. Frequent conference calls relayed the data on spinach purchases and sources to FDA, guiding the ongoing investigation of possible production sites of interest.

Between August 1 and October 6, a total of 199 persons infected with the outbreak strain of *E. coli* O157:H7 were reported to the CDC from 26 states. Among the ill persons, 102 were hospitalized, 31 had hemolytic (risking kidney failure), and 3 persons died. Eighty-five percent of patients reported illness onset from August 19 to September 5. Among the 130 patients for which a food consumption history was collected, 123 (95%) reported consuming uncooked fresh spinach during the 10 days before illness onset. In addition, *E. coli* O157:H7 with the same DNA matching the tainted strain was isolated from 11 open packages of fresh spinach that had been partially consumed by patients.

This outbreak strain of *E. coli* O157:H7 is one of 3,520 different *E. coli* O157:H7 patterns reported to CDC PulseNet since 1996. Infections with this strain have been reported sporadically to CDC's PulseNet since 2003, at an average of 21 cases per year from 2003 to 2005. This finding suggests that this strain has been present in the environment and food supply occasionally, although it had not been associated with a recognized outbreak in the past.

Parallel laboratory and epidemiologic investigations were crucial in identifying the source of this outbreak. Rapid collection of standard case exposure information by epidemiologists in affected states led to rapid identification of the suspected food source and public health alerts.

POTENTIAL FOR FUTURE CONTROVERSY

The farmer and cowboy are classical figures in U.S. history and culture. However, they do not fit well with modern industrialization of agriculture. Modern agribusiness is a group of large powerful corporations and banks. One battleground in this controversy is the cultural clash of old and traditional cultures with new ways of producing food.

The industrialization of agricultural practices is not new. Agricultural research at U.S. land grant colleges and universities helped to create the green revolution. The green revolution modernized many agricultural practices and

WHAT IS *E. COLI*?

E. coli is the abbreviated name of the bacterium in the Family Enterobacteriaceae named *Escherichia* (genus) *coli* (species). Approximately 0.1 percent of the total bacteria within an adult's intestines (on a Western diet) is represented by *E. coli*. although, in a newborn infant's intestines *E. coli*, along with lactobacilli and enterococci, represent the most abundant bacterial flora. The presence of *E. coli* and other kinds of bacteria within our intestines is necessary for us to develop and operate properly and to remain healthy. The fetus of an animal is completely sterile. Immediately after birth, however, the newborn acquires all kinds of different bacteria that live symbiotically with the newborn and throughout the individual's life. A rare strain of *E. coli* is *E. coli* O157:H7, a member of the EHEC—enterohemorrhagic *E. coli* group. Enterohemorrhagic means that it causes internal bleeding.

This strain of *E. coli* and all of its progeny produce a toxin. The toxin is a protein that causes severe damage to intestinal cells on the wall of the intestine. Internal bleeding occurs if left untreated and could lead to complications and death.

---

pushed them into greater productivity. Now some of the long-term results are more evident. Many people were fed. But some of the long-term consequences may be risky to public health and the environment. This is a large battleground in this controversy. These practices may also not be sustainable, especially with rapid climate change and urban population increases. This is an emerging battleground. The technological modernization of food production is continuing. Food can now be produced without soil. Although only in the research stage, the implications for food production are enormous. Food production would no longer have to be tied to the land. The new change in technology will face the same battlegrounds. It could also open environmental possibilities for former farmland.

See also Climate Change; Cumulative Emissions, Impacts, and Risks; Farmworkers and Environmental Justice; Genetically Modified Food; Organic Farming; Pesticides; Sustainability

Web Resources

Lettington, Robert J. L. "A Place for Agriculture at the Trade and Environment Table." Available at www.ictsd.org/dlogue/2001-07-30/Lettington.pdf. Accessed January 21, 2008.

Plant Physiology. "Agricultural Ethics." Available at www.plantphysiol.org/cgi/content/full/132/1/4. Accessed January 21, 2008.

U.S. Department of Agriculture. 1992. Societal Impacts of Adoption of Alternative Agricultural Practices. Available at www.nal.usda.gov/afsic/AFSIC_pubs/qb93–01.htm. Accessed January 21, 2008.

Further Reading: Douglass, Gordon K. 1984. *Agricultural Sustainability in a Changing World Order.* Boulder, CO: Westview Press; Gliessman, Stephen R., Eric W. Engles, and Robin Krieger. 1998. *Agroecology: Ecological Processes in Sustainable Agriculture.* Singapore: CRC Press; Kimbrell, Andrew. 2002. *The Fatal Harvest Reader: The Tragedy of Industrial Agriculture.* MO: Island Press; Loomis, R. S., and D. J. Connor. 1992. *Crop Ecology: Productivity and Management in Agricultural Systems.* Cambridge: Cambridge

University Press; Manno, Jack P. 2000. *Privileged Goods: Commoditization and Its Impact on Environment and Society.* Singapore: CRC Press; Wojcik, Jan. 1989. *The Arguments of Agriculture: A Casebook in Contemporary Agricultural Controversy.* Ashland, OH: Purdue University Press.

INDUSTRIAL FEEDING OPERATIONS FOR ANIMALS

Animal agriculture is switching to industrial practices to meet the needs of a growing world population and increase profits by replacing small to midsize animal farms with large, industrial-scale animal feeding operations (AFOs) that maximize the number of livestock confined per acre of land. Confinement of large numbers of animals in such operations can result in large discharges of animal feed- and waste-related substances (animal residuals) to the environment. The implications of waste management practices at AFOs for ecosystem viability and human health are very controversial. Potential effects of AFOs on the quality of surface water, groundwater, and air and on human health pose controversial issues.

CATTLE

Cattle, sheep, hogs, goats, and other animals have been raised for food all over the world for many years. Their environmental impacts are different based on the animal and the particular environment. Goats and hogs can have big impacts on ground cover and do long-term damage to sensitive ecotones like mountains. Cattle take large amounts of grassland to grow to market maturation. Environmentalists often object to eating beef because the environmental footprint of cattle raising is so large. Some have argued that rain forest deforestation from slash-and-burn techniques is motivated by a desire to expand grazing ranges for cattle. Cattle production is big business. There are about 500,000 concentrated animal feeding operations (CAFOs), about 20,000 of which are regulated under the pollution laws. Three states dominate feedlot cattle production: Texas, Kansas, and Nebraska account for two-thirds of all feedlot production of beef cattle in the United States. Ranchers are important political constituencies in these states around this issue.

Cattle feedlot operations are financially dominated by large corporations. Industrial feeding operations have refined the process of raising calves to slaughter-ready weight with industrial production methods. These focus on cost-to-profit measures and often prioritize size and weight gain, and time to market.

ENVIRONMENTAL IMPACTS

Large feedlot operations have provoked controversy in their communities, focused on the environmental damage caused by waste runoff and air pollution. Feedlot waste can be found in a watershed up to 300 miles away depending on the hydrology of that particular watershed.

MAD COW DISEASE AND OPRAH WINFREY

Oprah Winfrey Sued for Defaming Cattle in Texas

Mad cow disease, or bovine spongiform encephalopathy (BSE), is a fatal disease that affects the central nervous system of cattle. The United States does not import cattle from countries with reported cases of BSE nor do many other countries such as Japan.

Texas cattle ranchers sued TV talk-show host Oprah Winfrey for defamation in 1996. One of her guests stated that the cattle industry had potentially exposed Americans to mad cow disease by feeding cows the remains of live animals. The cattle ranchers requested money damages totaling $11 million.

After Oprah's alleged crime of airing a show examining mad cow risks in the United States, Texas's State Agriculture Commissioner Rick Perry asked the attorney general to use the state's new food-disparagement law to file a lawsuit against the Oprah show. When the attorney general declined, beef feedlot operator Paul Engler and a company named Cactus Feeders stepped in to shoulder the burden, hiring a powerhouse Los Angeles attorney to file a lawsuit that sought $2 million in damages plus punitive fines. "We're taking the Israeli action on this thing," Engler said. "Get in there and just blow the hell out of somebody." The lawsuit, filed on May 28, 1996, complained as follows: "The defendants allowed anti-meat activists to present biased, unsubstantiated, and irresponsible claims against beef, not only damaging the beef industry but also placing a tremendous amount of unwarranted fear in the public. . . . Defendants' conduct in making the statements contained herein and allowing those statements to be aired without verifying the accuracy of such statements goes beyond all possible bounds of decency and is utterly intolerable in a civilized society."

Oprah claimed victory after spending millions of dollars and years of her life battling the lawsuit. However, some feel that the real victors were Rick Perry and the cattle industry since they succeeded, as intended, in squelching news media coverage of mad cow risks in the United States, allowing to this day the continual feeding of hundreds of millions of pounds a year of cattle blood and fat to cattle, continuing the very practices that spread mad cow disease. Now Governor Perry presides over the first U.S. state to discover a home-grown case of the deadly animal and human dementia.

To this day, the real feed-ban firewall necessary to stop mad cow disease in the United States has not been constructed. Some contend that officials of the U.S. Department of Agriculture simply lie to the press and public when they say that a "ruminant to ruminant feed ban" prevents cattle protein from being fed to cattle in the United States, cutting off the spread of the disease. In reality, U.S. animal feed regulations allow hundreds of millions of pounds of cattle blood and fat to be fed back to cattle each year, including the widespread weaning of calves on cattle-blood protein in calf milk replacer and milk formula. In addition, one million tons a year of poultry litter is shoveled from barn floors at chicken factories and fed to cattle, although the spilled and defecated chicken feed in the litter can contain up to 30 percent mammalian meat and bone meal.

CAFOS IN OREGON

EPA's Concentrated Animal Feeding Operation (CAFO) Enforcement: The Case of Oregon

With an estimated 1.5 million head of cattle in Oregon, dairy and beef operations produce at least 7.5 million tons of manure per year that must be accounted for and kept out of Oregon's waters. Animal waste in water represents an environmental issue and a human health issue. For instance, animal waste is high in nutrients. When it enters a water body, oxygen can be depleted, preventing the breakdown of nutrients that can impact fish survival rates. Animal waste can also contain bacteria and viruses that are harmful to humans, including *E. coli* and *Salmonella*. Additionally, if cattle are allowed into streams, they can trample the streamside vegetation, which reduces shade cover and increases water temperature. It also increases erosion, that is, sediment deposition, which can severely impact the aquatic biota. A number of trout and salmon species found in Oregon are listed or have been proposed for listing as endangered species.

The Clean Water Act was enacted in 1972; however, some cases remain where CAFO owners have done little or nothing to keep animal waste from Oregon waters. CAFOs are defined as point sources under the NPDES program, and a discharge of animal waste to surface waters is illegal. These discharges often result from overflowing waste storage ponds, runoff from holding areas and concrete pads, or animals having direct access to surface waters. There have been many efforts to educate the CAFO owners by state and EPA regulators, yet violations persist. EPA's initial efforts began with dairies and have now included cattle feedlots as well as other CAFO operations, for example, hog farms, race tracks, and so on.

The direct involvement of the EPA in the regulation of CAFOs in Oregon is not new. In fact, the EPA has been involved in enforcing against dairy operations since 1994. The controversy that exists at this time is the result of the EPA expanding enforcement beyond the Oregon dairy industry to include beef cattle operations, which are a segment of the Oregon CAFO population that has received limited compliance inspections over the last several years, even though they have been subject to the regulations since 1972. During the past year 2007, the EPA directed some attention toward feedlots in eastern Oregon.

The EPA's objectives in taking federal enforcement actions against CAFOs in Oregon are as follows :

- Reduce the environmental and public health threat.
- Level the playing field among CAFOs by eliminating the economic advantage that violators have enjoyed over those who have invested capital to comply with the law.
- Encourage compliance and deter others from violating the law through education and public notice of penalties, thus supporting local efforts.
- Encourage the state of Oregon to reassume its lead role in CAFO enforcement.
- Use the authorities of the Clean Water Act as part of the salmon restoration efforts. The EPA is required by law to use these authorities under Section 7(a)1 of the Endangered Species Act.

The EPA has been directly involved in the regulation of Oregon CAFOs since 1994. In that time, the EPA has been involved in several activities that have resulted in the education of CAFO owners. These activities include the following:

1. Fact sheets, describing the EPA CAFO requirements and enforcement strategy, mailed annually to producers, assistance providers (NRCS), and industry associations such as the Oregon Dairy Association and the Oregon Cattlemen's Association.
2. Public meetings to discuss EPA requirements and enforcement strategy. These meetings were held in 1999–2000 in Pendleton, LaGrande, Enterprise, Baker City, Portland, Tillamook, Boise, and Tri-cities that were attended by both producers and assistance providers.
3. Several meetings with assistance providers (e.g., local conservation districts) to discuss EPA requirements and enforcement strategy.
4. Public notice of EPA enforcement actions against CAFO operations in Region 10.

What Types of CAFO Operations Are Being Enforced Against?

The EPA is targeting the worst cases first. In the cases filed to date, enforcement actions have been undertaken against operations with the confinement areas literally in the stream or where streams run directly through the confinement area with no attempt to keep manure out of the water and sites that have a direct discharge of waste to surface waters. Penalty assessments in Oregon have ranged from $11,000 to $50,000.

The EPA supports voluntary and community efforts to correct these problems, and EPA has supported many of these efforts with grant funds and people in the field. However, it is now the inspection year 2000–2001, and some recalcitrant operators remain who seem to need an incentive to do what others have done without enforcement actions being filed against them.

Lagoons are pools of water used to treat waste from animal feeding operations. They are an older, low-volume, low-cost waste treatment process but require maintenance. Waste treatment lagoons are often poorly maintained. They have broken, failed, or overflowed. They are prone to natural disasters like floods and hurricanes. When they overflow or break, the waste enters the watershed. Often the waste mixes with high levels of nitrogen and phosphorus from agricultural runoff. This can have major environmental impacts.

LAGOONS AND PUBLIC HEALTH

One major battleground of industrial feeding operations is the surrounding community. Gases are emitted by lagoons, including ammonia (a toxic form of nitrogen), hydrogen sulfide, and methane. These are all greenhouse gases and pollutants. The gases formed in the process of treating animal waste are toxic and potentially explosive.

Water contaminated by animal manure contributes to human diseases, potentially causing acute gastroenteritis, fever, kidney failure, and even death. According to the Natural Resources Defense Council, nitrates seeping from lagoons have contaminated groundwater used for human drinking water. Nitrate levels above 10 mg/L in drinking water increase the risk of methemoglominemia, or blue baby syndrome, which can cause deaths in infants.

THE LAGOONS HARM WATER QUALITY

There are also often cumulative effects from runoff within local watersheds because multiple large-scale feedlots cluster around slaughterhouses.

Watersheds far away are also affected by the atmospheric emission of gases from industrial feeding operations' lagoons, so that the environment is affected by both air and water pathways. Lagoons are often located close to water, which increases the potential of ecological damage. In many places, lagoons are permitted even where groundwater can be threatened. These communities have strong concerns, especially if they use well systems as many rural residents do. If water quantity is a local concern, then lagoons pose another battleground. The lagoon system depletes groundwater supplies by using large quantities of water to flush the manure into the lagoon. As water quantity decreases, pollutants and other chemicals become more concentrated. This decreases the quality of the remaining water dramatically.

THE HOG FARM CONTROVERSY

One of the biggest controversies over animal feeding operations occurred in South Carolina. Legislation introduced to accommodate hog-farming and hog-butchering operations created some of the controversy. Introduced under the title of a Right to Farm bill, the legislation passed the State House of Representatives without close scrutiny. Controversy began to build during the fall when it became clear the legislation would deprive local governments of some power to control land use. National media stories of environmental problems with large-scale hog farms in North Carolina started to get public attention. Those interested in economic development saw large-scale hog operations as a possible substitute for tobacco. Many of the objections to bringing the hog industry into South Carolina have to do with environmental degradation. One factor in the South Carolina hog controversy was how much waste the waterways could absorb. South Carolina water pollution permits have limited availability for waste. They did not want to use that remaining water-pollution capacity for low-return economic development.

South Carolina decided that if they lose the hog industry, they do not lose many economic benefits, and if they get it, it will come with difficult environmental problems that could hamper economic development over the long run.

EPA ATTEMPTS AT REGULATION

The U.S. Environmental Protection Agency (EPA) recognizes that animal feeding operations (AFOs) pose a variety of threats to human health and the environment. According to the EPA, pollutants from livestock operations include nutrients, organic matter, sediments, pathogens, heavy metals, hormones, antibiotics, dust, and ammonia. In response to increasing community complaints and the industrialization of the livestock industry, the EPA developed water-quality regulations that affect AFOs directly and indirectly, and it is a running battle. The focus of these actions is on the control of nutrient leaching and runoff. The development of this new set of rules is a large battleground.

Concentrated animal feeding operations (CAFOs) are defined as point sources under the Clean Water Act. They are required to obtain a permit to discharge treated and untreated waste into water. Effluent guidelines establish the best available technology economically achievable for CAFOs over a certain size threshold. A threshold is the maximum amount of a chemical allowed without a permit. Thresholds pervade U.S. environmental policy and allow industries that self-report their thresholds to escape environmental scrutiny. A constant regulatory battleground is lowering the threshold to expand the reach of the regulations to include all those with environmental impacts. Many communities and environmentalists complain that the thresholds for water discharges from industrial feeding operations are much too high, thereby allowing risky discharges into water. Industry wants to remain unregulated as much as possible because it perceives these regulations as decreasing profitability. The battleground about effluent thresholds for CAFOs is a major battleground. The new permitting regulations address smaller CAFOs and describe additional requirements such as monitoring and reporting.

HOW DO WE KEEP ENVIRONMENTAL IMPACTS FROM INDUSTRIALIZED FEEDING OPERATIONS FROM HARMING THE ENVIRONMENT?

Total Maximum Daily Load of Waste

The proposed total maximum daily load (TMDL) regulations and the development of nutrient water-quality criteria will impact AFOs indirectly. States are required to develop TMDLs for water bodies that do not meet the standards for nutrients or other pollutants. A TMDL is a calculation of the maximum amount of a pollutant that a water body can receive and still meet water-quality standards. Through the TMDL process, pollutant loads will be allocated among all permit holders. Animal feedlot operations may have to be slowed down if there is no room for their waste in the water. AFO management practices will be more strictly scrutinized in any event, creating a battleground for enforcement of environmental protection rules. This controversy will include the TMDL controversy when implemented at this level.

POTENTIAL FOR FUTURE CONTROVERSY

Industrial feed lot operations provide an efficient means of meat production. Communities and environmentalists are very concerned about their environmental impacts. They want to know more about these operations and usually ask for records on effluent discharges, monitoring systems for air and water, feed management, manure handling and storage, land application of manure, tillage, and riparian buffers. New federal regulations, growing population, community concern over environmental and public health impacts, and emerging environmental lawsuits are part of the battlefield for this controversy.

See also Cumulative Emissions, Impacts, and Risks; Fire; Industrial Agricultural Practices and the Environment; Rain Forests; Total Maximum Daily Loads (TMDL) of Chemicals in Water

Web Resources

The Legal Trend to Shift Liability for Environmental Damage. Available at www.rafiusa.org/programs/CONTRACTAG/Integrator_Liability.pdf. Accessed January 21, 2008.
Natural Resources Defense Council. America's Animal Factories: How States Fail to Prevent Pollution from Livestock Waste. Available at www.nrdc.org/water/pollution/factor/stwyo.asp. Accessed January 21, 2008.

Further Reading: Clay, Jason W. 2004. *World Agriculture and the Environment: A Commodity-by-Commodity Guide to Impacts and Practices.* Washington, DC: Island Press; MacLachlan, Ian. 2001. *Kill and Chill: Restructuring Canada's Beef Commodity Chain.* Toronto: University of Toronto Press; McNeely, Jeffrey A., and Sara J. Scherr. 2003. *Eco-agriculture: Strategies to Feed the World and Save Biodiversity.* Washington, DC: Island Press; Pfeffer, Ernst, and Alexander N. Hristov, eds. 2006. *Nitrogen and Phosphorus Nutrition of Cattle: Reducing the Environmental Impact of Cattle Operations.* Sydney, Australia: CABI Publishing; Yam, Philip. 2003. *The Pathological Protein: Mad Cow, Chronic Wasting, and Other Deadly Prion Diseases.* New York: Springer.

L

LAND POLLUTION

Controversies around land pollution abound. There are many types of waste and many governmental regulations at local, state, and federal levels. Waste sites, transfer stops, and routes all engender their own controversies. Communities do not want to be exposed to the risk waste may pose.

Waste is broadly defined as unwanted material left over from manufacturing processes or refuse from places of human or animal habitation. Within that category are many types of waste, including municipal solid waste, hazardous waste, and radioactive waste, which have properties that may make them dangerous or capable of having a harmful effect on human health and the environment. Waste and contaminated lands are particularly important to environmental health because they may expose land and living organisms to harmful material.

National, state, tribal, and local waste programs and policies aim to prevent pollution by reducing the generation of wastes at their source and by emphasizing prevention over management and subsequent disposal. Preventing pollution before it is generated and poses harm is often less costly than cleanup and remediation. Source reduction and recycling programs often can increase resource and energy efficiencies and thereby reduce pressure on the environment. When wastes are generated, the EPA, state environmental programs, and local municipalities work to reduce the risk of exposures. If land is contaminated, cleanup programs address the sites to prevent human exposure and groundwater contamination. Increased recycling protects land resources and extends the life span of disposal facilities.

HOW MUCH AND WHAT TYPES OF WASTE ARE GENERATED

The types of waste generated range from yard clippings to highly concentrated hazardous waste. Only three types of waste—municipal solid waste (MSW), hazardous waste (as defined by the Resource Conservation and Recovery Act [RCRA]), and radioactive waste—are tracked with any consistency on a national basis. Other types of waste, for which no or very limited national data exist, are not. This is a gaping hole is U.S. environmental policy. These other types of waste contribute a substantial amount to the total waste universe, although the exact percentage of the total that they represent is unknown.

Municipal solid waste, commonly known as trash or garbage, is one of the nation's most prevalent waste types. In 2000, the United States generated approximately 232 million tons of MSW, primarily in homes and workplaces—an increase of nearly 160 percent since 1960. During that time, the population increased 56 percent, and gross domestic product increased nearly 300 percent. In 2000, each person generated approximately 4.5 pounds of waste per day—or about 0.8 tons for the year—a per-capita increase from 2.7 pounds per day in 1960. For the last decade, per-capita waste generation has remained relatively constant, and the amount of MSW recovered (recycled or composted) increased more than 1,100 percent, from 5.6 million to 69.9 million tons in total. Combustion (incineration) is also used to reduce the volume of waste before disposal. Approximately 33.7 million tons (14.5 percent) of MSW were combusted in 2000. Of that amount, approximately 2.3 million tons were combusted for energy recovery.

The phrase *RCRA hazardous waste* applies to hazardous waste (waste that is ignitable, corrosive, reactive, or toxic) that is regulated under the RCRA. In 1999, the EPA estimated that 20,000 businesses generating large quantities—more than 2,200 pounds each per month—of hazardous waste collectively generated 40 million tons of RCRA hazardous waste. Comparisons of annual trends in hazardous waste generation are difficult because of changes in the types of data collected (e.g., exclusion of wastewater) over the past several years. But the amount of a specific set of priority toxic chemicals found in hazardous waste and tracked in the Toxics Release Inventory (TRI) is declining In 1999, approximately 69 percent of the RCRA hazardous waste was disposed of on land by one of four disposal methods: deep well/underground injection, landfill disposal, surface impoundment, or land treatment/application/farming.

In 2000, approximately 600,000 cubic meters of different types of radioactive waste were generated, and approximately 700,000 cubic meters were in storage awaiting disposal. By volume, the most prevalent types of radioactive waste are contaminated environmental media (i.e., soil, sediment, water, and sludge requiring cleanup or further assessment) and low-level waste. Both of these waste types typically have the lowest levels of radioactivity when measured by volume. Additional radioactive wastes in the form of spent nuclear fuel (2,467 metric tons of heavy metal) and high-level waste glass logs (1,201 canisters of vitrified high-level waste) are in storage awaiting long-term disposal.

WHAT IS THE EXTENT OF LAND USED FOR WASTE?

Between 1989 and 2000, the number of municipal landfills in the United States decreased substantially from 8,000 to 2,216. The combined capacity of all landfills, however, remained relatively constant because newer landfills typically have larger capacities. In 2000, municipal landfills received approximately 128 million pounds of MSW, or about 55 percent of what was generated. In addition to municipal landfills, the nation had 18,000 surface impoundments—ponds used to treat, store, or dispose of liquid waste—for nonhazardous industrial waste in 2000. Excluding wastewater, nearly 70 percent of the RCRA hazardous waste generated in 1999 was disposed of at one of the nation's RCRA treatment, storage, and land-disposal facilities. Of the 1,575 RCRA facilities, 1,049 are storage-only facilities. The remaining facilities perform one or more of several common management methods (e.g., deepwell/underground injection, metals recovery, incineration, landfill disposal).

The United States also uses other sites for waste management and disposal, but there are no comprehensive data sets that assess those additional sites or the extent of land now used nationally for waste management in general. Before the 1970s, waste was not subjected to today's legal requirements to reduce toxicity before disposal and was typically disposed of in open pits. Early land-disposal units that still pose threats to human health and the environment are considered to be contaminated lands and are subject to federal or state cleanup efforts.

WHAT IS THE EXTENT OF CONTAMINATED LAND?

Many of the contaminated sites that must be managed and cleaned up today are the result of historical contamination. Located throughout the country, contaminated sites vary tremendously. Some sites involve small, nontoxic spills or single leaking tanks, whereas others involve large acreages of potential contamination such as abandoned mine sites. To address the contamination, federal and state programs use a variety of laws and regulations to initiate, implement, and enforce cleanup. The contaminated sites are generally classified according to applicable program authorities, such as RCRA Corrective Action, Superfund, and state cleanup programs.

Although many states have data about contaminated sites within their boundaries, the total extent of contaminated land in the United States is unknown because few data are aggregated for the nation as a whole and acreage estimates are generally not available. A nationally accurate assessment would require both more detailed information on specific sites and consistent aggregation of those data nationally. To assess the full nature would require data on specific contaminants, as well as an assessment of risks, hazards, and potential for exposure to those contaminants.

The most toxic abandoned waste sites in the nation are listed on the Superfund National Priorities List (NPL). Thus, examining the NPL data provides an indication of the extent of the most significantly contaminated sites. NPL sites are located in every state and several territories. As of October 2002, there were 1,498 final or deleted NPL sites. An additional 62 sites were proposed to the NPL. When a proposed site meets the qualifications to be cleaned up under the Superfund program, it becomes a final NPL site. Sites are considered for deletion from the NPL list when cleanup is complete. Of the 1,498 sites, 846 sites are construction-completion sites, which are former toxic waste sites where physical construction for all cleanup actions is complete, all immediate threats have been addressed, and all long-term threats are under control. This is up from 149 construction completes in 1992.

The EPA also estimates that approximately 3,700 hazardous waste management sites may be subject to RCRA Corrective Action, which would provide for investigation, cleanup, and remediation of releases of hazardous waste and constituents. Contamination at the sites ranges from small spills that require soil cleanup to extensive contamination of soil, sediment, and groundwater. In addition, 1,714 of these 3,700 potential corrective action sites are high-priority sites that are targeted for immediate action by federal, state, and local agencies.

Other types of contaminated lands, for which data are very limited, include areas contaminated by leaking underground storage tanks and brownfields. Brownfields are lands on which hazardous substances, pollutants, or contaminants may be or have been present. Brownfields are often found in and around economically depressed neighborhoods. Cleaning up and redeveloping these lands can benefit surrounding communities by reducing health and environmental risks, creating more functional space, and improving economic conditions.

WHAT HUMAN HEALTH EFFECTS ARE ASSOCIATED WITH WASTE MANAGEMENT AND CONTAMINATED LANDS?

People who live, work, or are otherwise near contaminated lands and waste management areas are more vulnerable than others to the threats such areas might pose in the event of accident or unintended exposure to hazardous materials. Depending on factors such as management practices, the sources of contamination, and potential exposure, some waste, contaminated lands, and lands used for waste management pose a much greater risk to human health than others. Some areas, such as properly designed and managed waste management facilities, pose minimal risks.

Determining the relationship between types of sites and human health is usually complicated. For many types of cancer, understanding is limited by science and the fact that people usually are exposed to many possible cancer-causing substances throughout their lives. Isolating the contributions of exposure to contaminants to incidence of respiratory illness, cancer, and birth defects is extremely difficult—impossible in many cases. Nonetheless, it is important to gain a more concrete understanding of how the hazardous materials associated with waste and contaminated lands affect human populations.

Although some types of potential contaminants and waste are not generally hazardous to humans, other types can pose dangers to health if people are exposed. The number of substances that exist that can or do affect human health is unknown; however, the TRI program requires reporting of more than 650 chemicals and chemical categories that are known to be toxic to humans.

EPA's Superfund program has identified several sources of common contaminants, including commercial solvents, dry-cleaning agents, and chemicals. With chronic exposure, commercial solvents such as benzene may suppress bone marrow function, causing blood changes. Dry-cleaning agents and degreasers contain trichloroethane and trichloroethylene, which can cause fatigue, depression of the central nervous system, kidney changes (e.g., swelling, anemia), and liver changes (e.g., enlargement). Chemicals used in commercial and industrial manufacturing processes, such as arsenic, beryllium, cadmium, chromium, lead, and mercury, may cause various health problems. Long-term exposure to lead may cause permanent kidney and brain damage. Cadmium can cause kidney and lung disease. Chromium, beryllium, arsenic, and cadmium have been implicated as human carcinogens.

WHAT ECOLOGICAL EFFECTS ARE ASSOCIATED WITH WASTE AND CONTAMINATED LANDS?

Hazardous substances, whether present in waste, on lands used for waste management, or on contaminated land, can harm wildlife (e.g., cause major re-

---

CLEANUP OF THE EAGLE MINE SUPERFUND SITE

The Eagle Mine, southwest of Vail, Colorado, was used to mine gold, silver, lead, zinc, and copper between 1870 and 1984. After the mine closed, several contaminants, including lead, zinc, cadmium, arsenic, and manganese, were left behind, and they spread into nearby groundwater, the Eagle River, and the air, posing a risk to people and wildlife.

Colorado filed notice and claim in 1985 against the former mine owners for natural resource damages under Superfund. In June 1986, the site was placed on the National Priorities List, and shortly thereafter the state and the previous owners agreed to a plan of action. Cleanup operations included constructing a water treatment plant to collect mine seepage and other contaminated water sources; relocating all processed mine wastes and contaminated soils to one main, on-site tailings pile; capping that pile with a multilayer clean soil cap; and revegetating all disturbed areas with native plant species.

The water quality in the Eagle River began to show improvements in 1991; as zinc concentrations in the river dropped, the resident brown trout population grew. An October 2000 site review concluded that public health risks had been removed and that significant progress had been made in restoring the Eagle River. Today, biological monitoring is undertaken to evaluate the Eagle River's water quality, aquatic insects, and fish populations.

---

productive complications), destroy vegetation, contaminate air and water, and limit the ability of an ecosystem to survive. For example, if not properly managed, toxic residues from mining operations can be blown into nearby areas, affecting resident bird populations and the water on which they depend. Certain hazardous substances also have the potential to explode or cause fires, threatening both wildlife and human populations.

The negative effects of land contamination and occasionally of waste management on ecosystems occur after contaminants have been released on land (soil/sediment) or into the air or water.

POTENTIAL FOR FUTURE CONTROVERSY

The extent of land pollution is unknown at this time. Cleanup costs are enormous, which results in complex and expensive litigation to determine liability for these costs. In the United States, cities have only recently been included in the environmental protection policy umbrella. Controversies about land pollution generally focus on cleanup of the most serious wastes and/or relocation of the community. These controversies will increase.

See also Brownfields Development; Citizen Monitoring of Environmental Decisions; Cumulative Emissions, Impacts, and Risks; Ecosystem Risk Assessment; Industrial Agricultural Practices and the Environment; Sprawl; Toxics Release Inventory

Web Resources

CSIRO. Land Pollution. Available at www.csiro.au/csiro/channel/pch21.html. Accessed March 2, 2008.

Pollution. Land Pollution. Available at www.botany.uwc.ac.za/sci_ed/grade10/ecology/conservation/poll.htm#land. Accessed March 2, 2008.

Saving Our Environment. Dealing with Land and Water Pollution. Available at library.think-quest.org/C0111401/dealing_land_and_water_pollution.htm. Accessed March 2, 2008.

U.S. Environmental Protection Agency. National Priorities List (NPL). Available at www.epa.gov/superfund/sites/npl/status.htm. Accessed March 2, 2008.

Further Reading: Cairney, Thomas, and D. M. Hobson. 1998. *Contaminated Land: Problems and Solutions.* London: Spon Press; Genske, Dieter D. 2003. *Urban Land: Degradation, Investigation, Remediation.* New York: Springer; Nathanail, C. Paul, and R. Paul Bardos. 2004. *Reclamation of Contaminated Land.* New York: John Wiley and Sons; Owens, Susan E., and Richard Cowell. 2002. *Land and Limits: Interpreting Sustainability in the Planning Process.* New York: Routledge; Randolph, John. 2004. *Environmental Land Use Planning and Management.* Washington, DC: Island Press.

LAND-USE PLANNING IN THE UNITED STATES

Local land-use decisions are often at the heart of local environmental controversies. Local government makes most of these decisions. Residents are often surprised by the environmental impacts allowed. The ability of the federal government to preempt state governments, and the ability of state govern-

Industrial zones tend to be catchall zones with uses no one else wanted because of their risk. They include waste sites, chemical manufacturing plants, and former military bases. They are often the site of brownfields.

THE U.S. URBAN LANDSCAPE

Many of the early zoning schemes incorporated a doctrine of lesser included uses. A higher use could locate in a lower use, such as building a residence in a commercial area. This can sometimes explain a mix of uses in a given zone. Modern land-use zoning codes may intentionally have mixed-use zones, usually to decrease the trip generation of cars. This is an environmental consideration given the effect of vehicle emissions. Usually the primary consideration in enacting traffic laws is traffic flow and safety.

Since cities did not want to pay for the taking of private property, zoning proceeded slowly. It was not retroactively applied. Land uses that were in existence before zoning were generally allowed to continue until the property was sold or the use stopped. These are called *prior nonconforming uses.* Some of these uses can continue long after they are popular. Billboards can be one.

Special-use permits are given to property owners to allow otherwise prohibited uses of the land. Conceptually they cease when the use ceases or the owner changes. Variances are changes to the zoning that run with the land and continue indefinitely. They both generally increase the value of real property.

BATTLEGROUND: LAND-USE HEARINGS

Land-use hearings shape this battleground. Usually the local government makes these decisions. It can be a city council or a county board of supervisors. Sometimes, the applicant for a permit or variance must first go through a citizen planning commission. The decisions of these commissions are seldom binding but do create a record for any subsequent litigation. There are developers, teachers, consultants, professors, clergy, and sometimes environmentalists appointed to these commissions. These commissions will hold public hearings for citizens, often about environmental issues. It is in these hearings that many controversial environmental issues are first made public. After the citizen planning commission decision, it goes to the first elected level of government. They may also have hearings that can be focused on controversial environmental issues. Prior to litigation, the battleground for environmental land-use issues is developed in these types of hearings. It can be difficult to find out about these hearings. Notice and public participation in environmental decisions often start here and are controversies themselves.

The enforcement of land-use law tends to be very weak. There are no zoning police. Most zoning enforcement is complaint driven. City planners do not like neighborhood disputes to enmesh them in zoning disputes, although this occurs. High levels of citizen frustration often occur around environmental land-use issues. As noted above, a city can be preempted by the state to put a waste site in a given community. The citizens want action from their local government, and often there is not much they can do if they wanted to do so.

POTENTIAL FOR FUTURE CONTROVERSY

At the national level, the failure of local land-use law to fully consider environmental impacts apart from human impacts continues the idea that environmental impacts can be unlimited. The gulf between the political domination of land-use control and the need to control environmental impacts on the land is large.

Internationally, how human land-use controls affect global land cover is a big issue. Deforestation of tropical rain forests, overgrazing, and desertification are all land-use control issues. They also affect global systems of air and water.

There is a large potential for future controversy and for new environmental policy. As cumulative emissions, population, and waste all increase, and concern about sustainability mounts, citizens become more aware of their surroundings. They may be concerned about their health. Citizens, planning commissioners, and local elected officials face many environmental controversies and seek protection from environmental risks from their government. Land-use planning is one avenue for this.

See also Brownfields Development; Public Involvement and Participation in Environmental Decisions; Rain Forests; "Takings" of Private Property under the U.S. Constitution

Web Resources

National Environmental Health Association. Land Use Planning and Design. Available at www.neha.org/research/landuseplanning.html. Accessed March 2, 2008.
U.S. Environmental Protection Agency. Recommended EPA Land Use Resources. Available atoaspub.epa.gov/webimore/aboutepa.ebt4?search=16,200,867. Accessed March 2, 2008.

Further Reading: Freyfogle, Eric. 2003. *The Land We Share: Private Property and the Common Good.* Washington, DC: Island Press; Meltz, Robert, Dwight H. Merriam, and Richard M. Frank. 1999. *The Takings Issue: Constitutional Limits on Land-Use Control and Environmental Regulation.* Washington, DC: Island Press; Platt, Rutherford H. 2004. *Land Use and Society: Geography, Law, and Public Policy.* Washington, DC: Island Press; Randolph, John. 2004. *Environmental Land Use Planning and Management.* Washington, DC: Island Press; Silberstein, Jane, and Chris Maser. 2000. *Land-Use Planning for Sustainable Development.* Boca Raton, FL: CRC Press; Thomas, June Manning, and Marsha Ritzsdorf. 1997. *Urban Planning and the African American Community: Planning in the Shadows.* Thousand Oaks, CA: Sage Publications.

LANDSLIDES AND MUDSLIDES

Controversies around landslides and mudslides deal with issues of monitoring, evacuation, and human causes such as overdevelopment and logging. Liability for building in a landslide-prone area is a large legal controversy.

Mudslides and landslides can occur alone or with other natural disasters like hurricanes and floods. They can be human caused, as when an industrial mining-waste treatment lagoon bursts. They disrupt electric, water, sewer, and gas lines. Sewer, water, gas, and electric lines can also contribute to site destabilization, especially if there is any leakage on a slope composed mainly of soft material. They wash out roads and create health problems when sewage or floodwater spills down hillsides, often contaminating drinking water. Power lines and fallen tree limbs can be dangerous and can cause electric shock. Mudslides are also associated with volcanoes and earthquakes and can result in respiratory problems due to breathing of ash, fumes, heat, or gases. There is no question they are a natural disaster with large environmental impacts in the United States and in most other countries.

Landslides occur when masses of rock, earth, or debris move down a slope. Debris flows, also known as mudslides, are a common type of fast-moving landslide that tends to flow in channels. Landslides are caused by disturbances in the natural stability of a slope. They can accompany heavy rains or follow droughts, earthquakes, or volcanic eruptions. Mudslides develop when water rapidly accumulates in the ground and results in a surge of water-saturated rock, earth, and debris. Mudslides usually start on steep slopes and can be activated by other natural disasters. Areas where wildfires or human development have destroyed vegetation on slopes are particularly vulnerable to landslides during and after heavy rains.

HEALTH THREATS FROM LANDSLIDES AND DEBRIS FLOWS

In the United States, landslides and debris flows cause 25 to 50 deaths each year. The number is much higher worldwide. Many major world cities are surrounded by large communities of squatters. Uncounted, unregulated, densely packed, unsanitary, and often on defoliated, hilly landscapes, these communities are very vulnerable to landslides and mudslides. Health hazards associated with landslides and mudflows include:

- Rapidly moving water and debris that can lead to trauma
- Broken electrical, water, gas, and sewage lines that can result in injury or illness
- Disrupted roadways and railways that can endanger motorists and disrupt emergency services

Some areas are more likely to experience landslides or mudflows, including:

- Areas where wildfires or human modification of the land have destroyed vegetation
- Areas where landslides have occurred before
- Steep slopes and areas at the bottom of slopes or canyons
- Slopes that have been altered for construction of buildings and roads

(240 km) per hour. Such flows tend to follow valleys and are capable of knocking down and burning everything in their path. Lower-density pyroclastic flows, called *pyroclastic surges,* can easily overflow ridges hundreds of feet high. The climactic eruption of Mount St. Helens in the state of Washington on May 18, 1980, generated a series of explosions that formed a huge pyroclastic surge. This so-called lateral blast destroyed an area of 230 square miles (368 square km). Trees six feet in diameter were sheered down as far as 15 miles from the volcano. Volcano landslides range in size from small movements of loose debris on the surface of a volcano to a massive collapse of the entire summit or sides of a volcano. Steep volcanoes are susceptible to landslides because they are built partly of layers of loose volcanic rock fragments. Some rocks on volcanoes have also been changed to soft, slippery clay minerals by circulating hot, acidic groundwater.

LAHARS

Mudflows or debris flows composed mostly of volcanic materials on the sides of a volcano are called *lahars*. These flows of mud, rock, and water can rush down valleys and stream channels at speeds of 20 to 40 miles per hour and can travel more than 50 miles. Close to their source, these flows are powerful enough to rip up and carry trees, houses, and huge boulders miles downstream. Farther downstream they cover everything in their path with mud. They can occur both during an eruption and when a volcano is quiet. The water that creates lahars can come from melting snow and ice, torrential rainfall, or glacial and mountain lakes. Large lahars are a potential hazard to many communities downstream from glacier-clad volcanoes, especially under some climate change scenarios. Following the 1982 eruption of Mount St. Helens in the state of Washington, melting snow and ice triggered a lahar that traveled down the north flank of the mountain, following the channel of the North Fork of the Toutle River. Jammed with torn-up full-sized trees, it scoured the ground of everything.

HUMAN CAUSATION OF LANDSLIDES

Water, soil composition and bedrock, slope, vegetation, and previous land use, including road building, clear-cuts, and housing development, all play a role in creating a landslide risk. Some of the risk can be reduced. Poor management practices, such as clear-cutting and road building on steep slopes, can be avoided. Shoddy land-use planning practices, overdevelopment, and lack of environmental enforcement can increase the risk of landslides.

DOES LOGGING CAUSE LANDSLIDES?

One battleground is whether logging causes landslides. Most studies do find a higher incidence of slides in clear-cuts and logged areas than in unlogged forest. Some scientists question the methods used in the studies because they claim that landslides in uncut areas were undercounted because they were less visible. Others question the core geologic assumptions regarding slope stability. The differing results of landslide studies, and the general lack of clarity about how and why landslides occur, leads to a battleground. Those who wish to restrict activity on a slide-prone slope need to know which slopes are slide-prone. It is very difficult and expensive to accurately classify which slopes are slide-prone. Another difficult question and battleground is if logging, mining, and other road building and natural resource extraction activities do cause landslides, what is the policy about this? Should they be prevented from creating this risk, and if so, how? Should they be responsible for mitigation of any past risks created?

THE NATIONAL LANDSLIDE HAZARDS MITIGATION STRATEGY

The National Landslide Hazards Mitigation Strategy provides a framework for reducing losses from landslides and other ground failures. Although the strategy is national in scope, it is not exclusively federal or even exclusively governmental. Mitigation, defined as any sustained action taken to reduce and eliminate long-term risk to life and property, generally occurs at the state and local levels, and the strategy is based on partnerships with stakeholders at all levels of government and in the private sector.

The term *landslide* describes many types of downhill earth movements, ranging from rapidly moving catastrophic rock avalanches and debris flows in mountainous regions to more slowly-moving earth slides and other ground failures. In addition to the different types of landslides, the broader scope of ground failure includes subsidence, permafrost, and shrinking soils. The National Landslide Hazards Mitigation Strategy provides a framework that can be applied to other ground-failure hazards.

Landslides and other ground failures impose many direct and indirect costs on society. Direct costs include the actual damage sustained by buildings and property, ranging from the expense of cleanup and repair to replacement. Indirect costs are harder to measure and include business disruption, loss of tax revenues, reduced property values, loss of productivity, losses in tourism, and losses from litigation. The indirect costs often exceed the direct costs.

Much of the economic loss is borne by federal, state, and local agencies that are responsible for disaster assistance and highway maintenance and repair. Landslides have a significant adverse effect on infrastructure and threaten transportation corridors, fuel and energy conduits, and communications linkages. Ground-failure events have devastating economic effects on federal, state, local, and private roads, bridges, and tunnels every year. Railroads, pipelines, electric and telecommunication lines, dams, offshore oil and gas production facilities, port facilities, and waste repositories continually are affected by land movement. Road building and construction often exacerbate the landslide problem in hilly areas by altering the landscape, slopes, and drainages and by changing and channeling runoff, thereby increasing the potential for landslides. Landslides and other forms of ground failure also have adverse environmental consequences, such as dramatically increased soil erosion, siltation of streams and reservoirs, blockage of stream drainages, and loss of valuable watershed, grazing, and timber lands.

Before individuals and communities can reduce their risk from landslide hazards, they need to know the nature of the threat, its potential impact on them and their community, their options for reducing the risk or impact, and methods for carrying out specific mitigation measures. Achieving widespread public awareness of landslide hazards will enable communities and individuals to make informed decisions on where to live, purchase property, or locate a business. Local decision makers will know where to permit construction of residences, businesses, and critical facilities to reduce potential damage from landslide hazards.

DIFFERENT KINDS OF LANDSLIDES

Landslide is a general term for a wide variety of perceptible downward and outward movements of soil, rock, and vegetation under the influence of gravity. The materials may move by falling, toppling, sliding, spreading, or flowing. Some landslides are rapid, occurring in seconds, whereas others may take hours, weeks, or even longer to develop.

Although landslides usually occur on steep slopes, they also can occur in areas of low relief. Landslides can occur as ground failure of river bluffs, cut-and-fill failures accompanying highway and building excavations, collapse of mine-waste piles, and slope failures associated with quarries and open-pit mines. Underwater landslides usually involve areas of low relief and small slope gradients in lakes and reservoirs or in offshore marine settings. Landslides can be triggered by both natural changes in the environment and human activities. Inherent weaknesses in the rock or soil often combine with one or more triggering events, such as heavy rain, snowmelt, changes in groundwater level, or seismic or volcanic activity. Long-term climate change may result in an increase in precipitation and ground saturation and a rise in groundwater level, reducing the shear strength and increasing the weight of the soil. Erosion can remove the toe and lateral slope support, potentially causing landslides. Storms and sea-level rise often exacerbate coastal erosion and landslides. Earthquakes and volcanoes often trigger landslides.

The downward movement of land is generally classified as follows.

Falls—Abrupt movements of materials that become detached from steep slopes or cliffs, moving by free fall, bouncing, and rolling.

Flows—General term including many types of mass movement, such as creep, debris flow, debris avalanche, lahar, and mudflow.

Creep—Slow, steady downslope movement of soil or rock, often indicated by curved tree trunks, bent fences or retaining walls, and tilted poles or fences.

Debris flow—Rapid mass movement in which loose soils, rocks, and organic matter combine with entrained air and water to form a slurry that then flows downslope, usually associated with steep gullies.

Lahar—Mudflow or debris flow that originates on the slope of a volcano, usually triggered by heavy rainfall eroding volcanic deposits, sudden melting of snow and ice due to heat from volcanic vents, or the breakout of water from glaciers, crater lakes, or lakes dammed by volcanic eruptions.

Mudflow—Rapidly flowing mass of wet material that contains at least 50 percent sand, silt, and clay particles.

Lateral spreads—Often occur on very gentle slopes and result in nearly horizontal movement of earth materials. Lateral spreads usually are caused by liquefaction, where saturated sediments (usually sands and silts) are transformed from a solid into a liquefied state, usually triggered by an earthquake.

Slides—Many types of mass movement are included in the general term landslide. The two major types of landslides are rotational slides and translational landslides. In the rotational landslide the surface of rupture is curved concavely upward (spoon-shaped), and the slide movement is more or less rotational. A slump is an example of a small rotational landslide. In a translational landslide the mass of soil and rock moves out or down and outward with little rotational movement or backward tilting. Translational landslide material may range from loose, unconsolidated soils to extensive slabs of rock and may progress over great distances under certain conditions. Types of submarine and subaqueous landslides include rotational and translational landslides, debris flows and mudflows, and sand and silt liquefaction flows that occur principally or totally underwater in lakes and reservoirs or in coastal and offshore marine areas. The failure of underwater slopes can result from rapid sedimentation, methane gas in sediments, storm waves, current scour, or earthquake stresses. Subaqueous landslides pose problems for offshore and river engineering, jetties, piers, levees, offshore platforms and facilities, and pipelines and telecommunications cables.

Topple—A block of rock that tilts or rotates forward and falls, bounces, or rolls down the slope.

FEDERAL GOVERNMENT INVOLVEMENT

The U.S. Department of Agriculture Forest Service is a land-management agency with responsibility for natural resources in national forests. Most of the national forest lands are located in the mountainous areas of the western United States, including large parts of Alaska. The road system in national forests is comparable in size to state road systems. Consequently, designing low-volume roads to avoid landslide problems and repairing the damage to them from landslides are major tasks. Additionally, interstate and major state highways, railroad lines, oil and gas pipelines, and electric transmission corridors pass through the national forests. Assessing landslide hazards along such projects is increasingly important. National forests generally occupy the headwaters of major rivers, increasing the importance of watershed management, especially for those watersheds where anadromous fisheries and significant inland fisheries are present.

Increased landslide activity can produce sediment loads that degrade water quality and adversely affect fish habitat. Landslide hazard can be a more localized, but equally important, problem in national forests where development of large ski resorts, mines, or hydroelectric facilities takes place. Major wildfires can denude watersheds and lead to short-term landslide activity. The potential for loss of life and damage from debris flows initiated by precipitation events on burned watersheds must be considered in national forests, especially those having developed, private inholdings and adjacent urban areas.

A primary landslide hazard–related activity conducted by Forest Service personnel is evaluating landslide potential in environmental assessments or reviewing environmental assessments prepared by forestry project proponents. Environmental or engineering geologists, as one of their primary duties; minerals geologists, as a related duty; or other earth scientists, where geologists are unavailable, carry out these evaluations. Engineering geologists and geotechnical engineers carry out environmental assessments and participate in designs to address landslide hazard to system roads. Another activity is assessing damage from landslides following major natural disasters. The most formalized of these assessments is the Burned Area Emergency Rehabilitation procedure instituted during major wildfires. This activity also includes participating in development of stabilization and restoration projects to counter wildfire damage. A national geographic information system (GIS) network of national forest lands and a database that includes landslide information is under development. The landslide hazard information for this GIS is generated from U. S. Geological Survey (USGS) and State Geological Survey information and mapping by

FOREST SERVICE GEOLOGISTS

The National Oceanic and Atmospheric Administration (NOAA) and National Weather Service (NWS) are involved in landslide mitigation through their roles in the Federal Response Plan and its mission of providing services for the protection of life and property. The National Weather Service works with other federal, state, and local agencies by providing forecasts of hydrologic and meteorological conditions for landslide prediction and mitigation efforts. This assistance may include on-scene meteorological personnel to assist in emergency -response activities at landslides. The NOAA Weather Radio and other NWS dissemination systems broadcast civil emergency messages concerning landslide warnings and response and recovery efforts at the request of local, state, and federal emergency management officials.

U.S. Army Corps of Engineers' mission includes planning, designing, building, and operating water resources and civil projects in the areas of flood control, navigation, environmental quality, coastal protection, and disaster response, as well as the design and construction of facilities for the army, air force, and other federal agencies. In forming this broad mission, the corps has addressed a full range of technical challenges associated with landslides and ground failure. The Corps' engineering geologists, geotechnical engineers, and geophysicists have been involved in the assessment, monitoring and analysis, and mitigation of landslides in a wide range of settings at locations around the world, as well as basic and applied research on topics directly related to the analysis and mitigation of landslides and ground failures. Landslide assessment activities by corps scientists and engineers have included investigations of landslides of various mechanisms and scales along navigable waterways such as the Mississippi and Ohio rivers that result in serious navigation hazards and threats to or loss of flood-protection works.

Landslides also play an important role in the erosion of the shoreline; the protection of shoreline is a major responsibility of the corps. Many corps dam-site investigations have involved the identification and assessment of past and potential landslides. Corps have been involved in monitoring active landslides and ground failure in both natural and engineered soils and earth materials. These tasks have focused on identifying the temporal and spatial variability of earth movements and identifying causal factors. Monitoring data have been used along with detailed site information to analyze the stability of a landslide in terms of initial movements, present conditions, and conditions after mitigation actions. As an engineering agency, the corps has a significant role in the planning, design, and construction of landslide-mitigation measures associated with the protection of its civil and military projects. Specific methods for reducing landslide hazards and increasing slope stability have been developed and implemented by corps engineers at sites around the world.

The corps' role in initial engineering geological investigation, engineering analysis, remedial design, implementation, construction, and postproject monitoring is of particular value to the nation and the international community. The corps has an important national mission in disaster response. This mission has involved the corps in responding to landslides, especially those resulting from floods, hurricanes, volcanic eruptions, and earthquakes. The corps' role in international disaster response has become a major focus in landslide engineering. Recent landslide assessments, analysis, and mitigation efforts have been conducted in Venezuela, Honduras, Nicaragua, Colombia, Peru, Haiti, Puerto Rico, South Korea, and the Philippines. Research at the Corps Engineering Research and Development Center includes the development and testing of analytical tools and assessment methods and approaches for landslide mitigation. Basic research in soil and rock mechanics, geomorphology, hydrogeology, remote sensing, geophysics, and engineering geology has resulted in advancements in the understanding of the causative factors and mechanics of landslides and ground failures.

The Bureau of Land Management is a federal agency that manages multiple uses of approximately 264 million surface acres of federal land located primarily in 12 western states. A relatively small portion of this land is located in steep mountainous terrain with geologic and climatic conditions resulting in high landslide hazards, such as in western Oregon, northern California, and northern Idaho. Many landslides on public land are the result of natural disturbance events, but land-management activities, including road building, timber harvest, historic mining, and water impoundments, can contribute to their occurrence. The Bureau of Land Management does not have an agency-wide landslide hazards program or specialized personnel. Landslide hazards prevention activities conducted by the bureau's local field offices include identification of unstable slopes by using aerial photograph interpretation, landslide hazards guides, on-site indicators, predictive models, and limited inventory and monitoring of landslides. Prevention and mitigation of landslides are accomplished by using a variety of methods. Existing roads may be closed and obliterated, rerouted, or kept open and stabilized with additional runoff-control structures, subsurface drainage control, or other techniques. Routine road maintenance is an important factor in helping to reduce landslide hazards. Prudent route analysis and design to minimize landslide hazard are employed for new roads in landslide-prone areas. Hazardous-fuels management can reduce the risk of catastrophic wildfires that could increase landslide hazards. Timber management silvicultural practices are employed to maintain root strength where needed for slope stability. Sites that are a threat to human health and safety, roads and recreational facilities, water quality, fisheries and aquatic habitat, and other valued resource are stabilized, and sediment is controlled with revegetation and structural means.

Many national parks are geologically active, exposing park visitors, staff, and infrastructure to geologic hazards. Landslides, including slope failures, mudflows, and rockfalls, adversely affect parks, causing deaths and injuries, closing roads and trails, and damaging park infrastructure. Recent examples include several rockfalls in Yosemite Valley, each resulting in one fatality; damaging landslides in Shenandoah National Park triggered by torrential rains; repeated slope failures fed by artificial aquifers at Hagerman Fossil Beds National Monument; landslides that closed roads in Zion and Yellowstone National Parks; and the threat of large debris flows at Mt. Rainier. USGS scientists have provided insights essential to effective response to landslide hazards at these and other national parks. Because it is a natural process, landslide activity is generally allowed to proceed unimpeded in national parks unless safety is a concern.

However, where people have destabilized the landscape (e.g., by logging, mining, and road building), disturbed lands are restored where practical to their predisturbance condition. To reduce risk from landslides and other geologic hazards, park planners must incorporate information from hazard assessments and maps into decisions about appropriate sites for facilities such as campgrounds, visitor centers, and concession areas. Planners face difficult choices as they attempt to balance risks from different hazards, such as floods and rockfalls in confined valleys, with at the same time providing public access to popular but potentially hazardous areas. When a landslide or other hazard occurs, park personnel must quickly rescue people, stabilize structures, and clear debris from roads and other public areas. Then park personnel must work with experts to assess the nature and extent of the event and the risk of recurrence. Short-term studies are required to help managers decide whether and when to reopen affected areas; then more detailed research is often needed to make informed decisions about future use of the immediately affected area and other areas that may face similar hazards.

POTENTIAL FOR FUTURE CONTROVERSY

Given current land-development trends and land-use planning processes, it is likely that more controversies about landslide liability will develop. Many currently feel it is an individual responsibility to know whether a site is a landslide hazard. Others point out that in fact most people do know whether they are in a landslide risk area or not. State intervention is necessary, they argue, because people need protection.

The scientific information around landslide monitoring and assessment has grown rapidly as this controversy has heated up. Arguments about whether a specific act of clear-cutting timber or building a ski highway on a steep mountain caused a specific landslide will be in courts and state agencies for years to come. Because many factors can cause a landslide, and because landslides can occur over a long time, this will be difficult to prove in court. However, the new wave of scientific information will inform state and local policy makers empowered with the authority to make environmental decisions.

See also Climate Change; Environmental Vulnerability of Urban Areas; Evacuation Planning for Natural Disasters; Federal Environmental Land Use; Logging; Mining of Natural Resources; Sprawl; Watershed Protection and Soil Conservation

Web Resources

Forecasting the Collapse and Runout of Giant Catastrophic Landslides. Available at www.benfieldhrc.org/landslides/Landslides.pdf. Accessed January 21, 2008.

Guidelines on Logging Practices for the Hill Forest of Peninsular Malaysia. Available at www.fao.org/docrep/W3646E/w3646e0d.htm. Accessed January 21, 2008.

Logging and Landslides: A Clear-Cut Controversy. Available at www.wildfirenews.com/forests/forest/index.html. Accessed January 21, 2008.

Further Reading: Baldwin, John H. 1985. *Environmental Planning and Management.* Boulder, CO: Westview Press; Glade, Thomas, ed. 2005. *Landslide Hazard and Risk.* New York: John Wiley and Sons; MacGuire, Bill, Ian M. Mason, and Christopher R. J. Kilburn. 2002. *Natural Hazards and Environmental Change.* New York: Oxford University Press; Matthias, Jakob O. 2005. *Debris-Flow Hazards and Related Phenomena.* New York: Springer; Rybář, Jan, Josef Stemberk, and Peter Wagner. 2002. *Landslides: Proceedings of the First European Conference on Landslides, Prague, Czech Republic.* Oxford: Taylor Francis; Stoltman, Joseph P., John Lidstone, and Lisa M. Dechano, eds. 2005. *International Perspectives on Natural Disasters: Occurrence, Mitigation, and Consequences.* New York: Springer.

LEAD EXPOSURE

Environmental exposure to lead can cause life-impairing if not deadly consequences. Lead is a metal that can harm humans and the environment and is present in many pollutants and consumer goods. Many of the controversial issues surrounding lead exposure relate to scientific disagreement over the extent and severity of exposure. Other controversies relate to political disagreements over who is exposed, when to test for lead exposure, and who will pay for diagnosis and treatment.

WHAT IS LEAD?

Lead is essentially a mineral. Minerals are formed by hot solutions working their way up from deep below the earth's rocky crust, and crystallizing as they cool near the surface. From early history to the present, mining lead is part of human culture. Lead was a key component in cosmetic decoration as a pigment; a spermicide for informal birth control; a sweet-and-sour condiment popular for seasoning and adulterating food; a wine preservative for stopping fermentation or disguising inferior vintages; the malleable and inexpensive ingredient in pewter cups, plates, pitchers, pots and pans, and other household artifacts; the basic component of lead coins; and the material of children's toys. Most important of all was lead's suitability as inexpensive and reliable water pipes. Lead pipes are still used in many parts of the world and are still present in the United States. Lead pipes kept the Roman Empire supplied with water. From then until now lead has been part of our society.

Human exposure to lead is a serious public health problem. Lead adversely affects the nervous, hematopoietic, endocrine, renal, and reproductive systems of the body. Of particular concern are the effects of relatively low levels of lead exposure on the cognitive development of children. Since the 1970s, federal environmental regulations and abatement efforts have mainly focused on reducing the amount of lead in gasoline, paint, and soldered cans. In addition, some federal programs have supported screening for lead poisoning in children by state and local health departments and physicians, as well as lead abatement in housing. Currently, lead exposure usually results from lead in deteriorating household paint, lead at the workplace, lead used in hobbies, lead in some folk medicines and cosmetics, lead in children's toys, and lead in crystal or ceramic containers that leaches into water or food.

Since the late 1970s, the extent of lead exposure in the U.S. population has been assessed by the National Health and Nutrition Examination Surveys (NHANES). These national surveys have measured blood lead levels (BPb) of tens of thousands of children and adults and assessed the extent of lead exposure in the civilian population by age, sex, race/ethnicity, income, and degree of urbanization. The surveys have demonstrated an overall decline in BPbs since the 1970s, but they also have shown that a large number of children continue to have elevated blood lead levels (10 µg/dl). The U.S. EPA claims that due to environmental regulations, airborne lead amounts have decreased by 90 percent from 1980 to 2005.

Sociodemographic factors associated with higher blood lead levels in children were non-Hispanic black race/ethnicity, low income, and residence in older housing. The prevalence of elevated blood lead levels was 21.9 percent among non-Hispanic black children living in homes built before 1946 and 16.4 percent among children in low-income families who lived in homes built before 1946. Overall, blood lead levels continue to decline in the U.S. population. The disproportionate impact on urban people of color makes it an environmental justice issue, although lead can work its way through ecosystems and affect other groups downwind or downstream.

Lead is a highly toxic metal that was used for many years in products found in and around our homes, schools, and workplaces. Lead can cause a range of health effects, from behavioral problems and learning disabilities, to seizures and death. Children six years old and under are most at risk, because their bodies are growing quickly.

The EPA is playing a major role in addressing these residential lead hazards. In 1978, there were nearly three to four million children with elevated blood lead levels in the United States. By 2002, that number had dropped to 310,000 children, and it continues to decline. Since the 1980s, the EPA has phased out lead in gasoline, reduced lead in drinking water and in industrial air pollution, and banned or limited lead used in consumer products, including residential paint. States and cities have set up some programs to identify and treat lead-poisoned children and to rehabilitate deteriorated housing. Parents have greatly helped to reduce their children's lead exposure by eliminating lead in their homes, having their children's blood lead levels checked, and learning about the risks of lead for children.

Some population groups continue to be at disproportionately high risk for elevated lead exposure. In general, these are people with low income, people of non-Hispanic black race, and people who live in older housing. Residence in a central city with a population of more than one million people was also found to be a risk factor for lead exposure.

These high-risk population groups are important to recognize for targeting of public health efforts in lead-poisoning prevention. Leaded paint, especially in older homes, is a continuing source of lead exposure. In the United States, approximately 83 percent of privately owned housing units and 86 percent of public housing units built before 1980 contain some lead-based paint. Commercial and industrial structures often have much more lead paint. Bridges are often constantly being painted and use large amounts of lead paint. The areas under the bridges are often contaminated with lead. If the area under a bridge is land, the lead accumulates on the ground and in the dirt. Many bridges of this type are older and in urban areas. Between 5 and 15 million homes contain paint that has deteriorated to the point of being a health hazard. Lead hazard control and abatement costs are highly variable, depending on the extent of the intervention, existing market conditions, type of housing, and associated housing rehabilitation work. Because of the high cost of abatement, the scarcity of adequately trained lead-abatement professionals, and the absence (until 1995) of federal guidelines for implementing less-costly methods of leaded paint hazard containment, residential lead-paint-abatement efforts have focused on homes in which there is a resident child with an elevated BPb, rather than on homes that have the potential to expose a child to lead. Similarly, publicly funded lead-poisoning prevention programs have focused on screening children to identify those who already have elevated BPbs so that they may receive interventions, rather than on preventing future lead exposure among children without elevated levels. This is a battleground in the public health arena. At least some of the adverse health effects that occur even at relatively low levels of lead may be irreversible. Lead exposure prevention efforts through screening are important to ensure that children with elevated BPbs receive prompt and effective interventions to reduce further lead exposure and minimize health consequences. These types of programs are not consistently well funded.

One source of lead in many countries is vehicle emissions. In the United States these emissions are much lower because of leadfree gasoline in cars. However, other airborne vehicle emissions such as from diesel may have a detrimental effect on children. In urban areas near congested roads, exposure to lead via air may be a large exposure vector.

A controversial issue is whether ambient air quality standards are adequate to protect the health of children. Currently a few state environmental agencies are working to identify toxic air contaminants that may cause infants and children to be especially susceptible to risks. In many cases, children may have greater exposure than adults to airborne pollutants. Children are often more susceptible to the health effects of air pollution because their immune systems and developing organs are still growing. Lead that is inhaled is more easily lodged in the fast-growing bones of children. It may take less exposure to airborne lead to initiate an asthma attack or other breathing ailment due to the smaller size and greater sensitivity of a child's developing respiratory system.

POTENTIAL FOR FUTURE CONTROVERSY

As a pervasive material in civilization, lead is ubiquitous. Lead in the water pipes and drinking vessels may have poisoned the people of Rome and contributed to its downfall. Lead is still in plumbing systems throughout the United States. In Washington, D.C., some of the leaded sewer and water pipes dead end. This means that there is no way to flush them clear of corrosive wastes and debris. This increases the rate of corrosion of these pipes and leaches lead into the water. When the pipe breaks it may slowly seep lead-contaminated water into the surrounding ground. The ground could be anywhere the pipe is found, near a road, river, school, factory, or any place connected to or near the break. This is a matter of great concern in chemically polluted older communities and can develop into a land-use battleground or a battleground around the environmental permits necessary, if any. Refining methods to assess health risks from lead that may exist at proposed and existing school sites is under way. Because lead is found in so many places where vulnerable populations exist, from children's toys to hospitals, controversy around lead exposure and cleanup will increase.

The disposal of lead as hazardous waste is also a controversial issue. Demolition of lead contaminated houses, factories, and infrastructure creates hazardous waste. If burned in an incinerator the lead could be spread in the air as particulate matter. The disposal of this type of toxic ash can also become a controversy. If the lead-contaminated structure simply stays in place, the lead can affect the soil and nearby water. The reluctance to expend the resources necessary to protect the most exposed people from lead contamination ensures a continuing controversy. This controversy may not necessarily diminish even as science removes uncertainty about lead contamination.

See also Air Pollution; Cumulative Emissions, Impacts, and Risks; Human Health Risk Assessment; Mining of Natural Resources; Toxics Release Inventory; Transportation and the Environment

Web Resources

Department of Health and Human Services Agency for Toxic Substances and Disease Registry. Lead Toxicity Cover Page. Available at www.atsdr.cdc.gov/HEC/CSEM/lead/expo sure_pathways.html. Accessed January 21, 2008.

Further Reading: Breen, Joseph J., and Cindy R. Stroup. 1995. *Lead Poisoning: Exposure, Abatement, Regulation.* Singapore: CRC Press; Centers for Disease Control. 1991. *Preventing Lead Poisoning in Young Children: A Statement by the Centers for Disease Control.* Atlanta: Centers for Disease Control; Moore, Colleen F. 2003. *Silent Scourge: Children, Pollution, and Why Scientists Disagree.* New York: Oxford University Press.

LITIGATION OF ENVIRONMENTAL DISPUTES

The use of litigation to resolve environmental policy disputes is itself controversial. A hybrid type of lawsuit is created by environmental citizen suits, which allow environmental legal defense groups to bring certain kinds of environmental disputes into federal courts. The prevailing party gets attorneys' fees. Industry strongly resists most laws allowing citizen suits, and many frustrated communities and almost all environmentalists support them.

The use of courts to make environmental policy is a sign of how important this is to society. Courts give the most final of judgments in the United States, with the force of law behind them. Courts are also one way citizens can petition their government with grievances, but federal court jurisdiction is very limited and expensive. Federal judges are appointed for life by the president, with the advice and consent of the U.S. Senate. This process is very political and controversial itself.

Courts are also very limited in their ability and expertise to resolve environmental disputes. They are limited to the cases before them, which are limited to two parties to the dispute. Occasionally, a third party may enter as an intervener. At issue in a given case are the facts, the issues, the law applied to them, and the judgment. The environment is not directly represented in any way. Contiguous property owners may never know what judicial actions are occurring in regard to the environment. There are some emerging exceptions that will be battlegrounds. In New Mexico, a third party outside the sale of land from a buyer to a seller can challenge the transaction on environmental grounds. State and noninvolved federal agencies, other communities, or anyone else generally will not know anything about the case if it is settled because industry generally settles on the condition that it remain undisclosed and that no admission of any wrongdoing is made. Many environmentalists feel the federal courts are stacked in favor of industry because federal judges are chosen from larger law firms and other conservative groups. Also, the confidential aspect of judicial decision making is generally not transparent enough to provide a sound basis for environmental policy. Federal judges are known to seal records and depositions without notice, examine evidence in camera, and give decisions with opinions on the express provision that they not be used as precedent. Some federal jurisdictions now expressly do not allow their decisions to be used as precedent. While the implications of this are deeply controversial, its net effect on environmental policy is to make judicial environmental decisions inconsistent and political. Judicial decision making is, however, relied upon for setting the parameters of U.S. environmental policy, and therefore sets some of prominent contours of this battlefield.

In the 1970s, when the EPA was formed, several environmental laws had powerful citizen suit provisions attached. These basically allow environmental defense groups to sue for citizens if the EPA is not enforcing water and air laws. The EPA and industrial site must receive a letter of intent to sue if the EPA does not enforce the law at least 60 days before the action is filed in court. If the case goes forward, the prevailing side gets attorneys' fees. These have been very important cases that forced the EPA, then a new and small federal agency, to move quickly in advancing a defensible legal basis for environmental protection. Without environmental groups advocating in court the basic standards for clean air would have taken much longer to develop, as one example. However, lawsuits also engender ill will that can last after a court makes an environmental decision. No matter what any court decides, the environment will always act independently. Unlike many other judicial decisions, environmental decisions can be measured by a standard of environmental condition to determine the soundness of that judicial decision on environmental policy. Many lawsuits could have been classified as environmental before passage of the Clean Air Act and the formation of the EPA. Without clear statutory support courts were reluctant to intervene in land-use or environmental decisions and often denied standing for such suits for failing to state a recognizable cause of action. Without judicial intervention people lacked a legal remedy for environmental or land-use cases. The environment continued to worsen until the states and the federal government reacted to the will of Congress and formed environmental agencies and passed environmental laws that clearly gave citizens the right to sue in federal court.

POTENTIAL CITIZEN SUIT EXPANSION

On January 12, 2000, the U.S. Supreme Court decided *Friends of the Earth, Inc. v. Laidlaw Environmental Services, Inc.* The 7–2 opinion, delivered by Justice Ginsburg, marked a reversal toward a narrow reading of standing requirements in environmental citizen suits. This decision is important to citizen suits. Friends of the Earth and other plaintiff-petitioners filed a citizen suit against Laidlaw Environmental Services in 1992, alleging noncompliance with the water permit at Laidlaw's wastewater treatment plant in Roebuck, South Carolina. In 1997, the district court had assessed a civil penalty of over $400,000 against Laidlaw. On appeal, the court held that since Laidlaw had come into compliance with its permit before the judgment by the district court, the case had become moot. That meant no penalty could be assessed. The case was appealed to the Supreme Court. The Court reversed the Fourth Circuit's decision, rejecting Laidlaw's mootness claim and allowing the civil penalty to stay. This landmark decision encourages future citizen suits. In the majority opinion of the Court, plaintiffs' concern that water was polluted, and a belief that the pollution had reduced the value of their homes, was deemed sufficient for standing.

PRECLUSION OF CITIZEN SUITS AND GOVERNMENT ENFORCEMENT ACTIVITY

Because the object of citizen-suit provisions is increased levels of enforcement, citizen suits may be filed as long as an enforcement action is not already under way by the relevant agency. That a citizen suit is under way, however, does not necessarily preclude an agency enforcement action. The question many have is why allow an agency action to follow a citizen suit and not also the other way around? Citizen lawsuits allows plaintiffs to prevail for discovering and initiating suits based on unpunished, current violations only. It is a complex question that forms one of the many layers in the battleground on the use of litigation to resolve environmental disputes.

ATTORNEYS' FEES

One controversial area of the use of litigation to solve environmental issues is the battleground of attorneys' fees. The prevailing party generally gets attorneys' fees. The rationale is that the environmental laws are premised on citizen suits and their use in enforcement. In litigation-intensive cases with many appeals, attorney fees can be expensive because of the long hours. In some of the endangered species litigation around the spotted owl in Oregon, the attorneys, Western Environmental Law Center, were awarded one million dollars in attorneys' fees. This reimbursement can be helpful for plaintiffs, because rates are based on the market rate for for-profit, private attorneys, not the rates charged by the public-interest lawyers such plaintiffs typically use. Both government and industry generally have their own in-house lawyers.

AN ENVIRONMENTAL JUSTICE CITIZEN SUIT?

U.S. environmental law is a relatively new field; other branches of law have historically been used to remedy environmental problems. Nuisance actions were the most popular, because they allow a successful claimant not only to receive compensation but also a court order to abate the nuisance, such as a smell or smoke. Historically, however, tort law, based as it is on the protection of individual rights and the need to prove specific injury, has not been a significant means of preventing environmental degradation.

It was, and is, difficult to access judicial resources. Decisions were unpredictable and sometimes took years. For many years of U.S. history, courts were for the privileged. Generally one had to own property and be over 21 years old, white, and male. Courts were not good places to decide environmental issues that spread across many classes and races of people. The litigation convinced local and state governments to adopt measures to tackle the most pressing environmental problems. Most of these early measures were, in fact, enacted after sporadic crises that endangered public health.

Modern Environmental Law

Since the 1970s, environmental law has experienced an unprecedented growth in many countries. This was made possible through the enactment of new statutes and regulations that provide for higher standards of environmental protection. The level of government that has enacted these instruments varies from one country to another. In the United States, the federal government has adopted most of the important environmental statutes, but their implementation is delegated to the states through a complex system of incentives and responsibilities. Citizens can sue the EPA if the EPA is failing to react to violations of environmental law.

Environmental Citizen Suits in the United States

Currently, sixteen states have environmental citizen-suit statutes on the books. In general, these statutes give citizens, or any person, the right to sue the state, a private party, or both, to protect the state's environment. Some citizen-suit statutes provide only for injunctive relief to stop harmful activity or to force the state to act, while others authorize the award of money damages as well. Whether the citizen filing the lawsuit can recover her or his attorneys' fees and litigation costs is another variation, and battleground, among the statutes. Finally, different meanings for the term environment exist. Almost all of the states with citizen-suit statutes allow suits against any party, though some are limited to actions against only the state. Half of the statutes do not require a violation of law for a suit to be filed. Half allow equitable relief only, while two others also provide for penalties and money damages, and the rest are silent on this issue. Only a very few provide for costs to the prevailing party. Finally, only two states actually define the words *environment* or *natural resources*.

Federal Citizen-Suit Provisions

Nearly every major environmental statute provides for citizen suits. In general, these provisions allow any citizen to sue any person to enforce the specific requirements or limitations of the environmental protection statute.

Questions and Answers about Using Citizen Suits to Achieve Environmental Justice

There has been substantial legislative activity around the failure of the EPA to obey presidential executive orders on environmental justice. Executive Order 12898 is now reported out of the House of Representatives Environment Committees as a proposed law, specifically because the EPA refuses to obey this order in spite of numerous requirements and decisions. During legislative caucusing around this bill the issue of an environmental justice citizen was debated, but ultimately not included. Some of the questions that went back and forth during these debates include the following.

When Do We Need Citizens to Enforce the Law?

When government cannot be relied on to enforce the law. The EPA has shown this time and again with its resistance to complying with the Office of Inspector General reports, which are minimal and required by law.

Some have argued that a citizen-suit provision is needed to bring cities into the national environmental policy arena in order to comply with increasing global demands for clean air. The United States has only recently incorporated cities into environmental regulations, yet that is where most of the pollution and people exist. Global warming and climate change concerns among other nations and in some industries will require the United States to enforce stricter environmental laws in all areas. Citizen suits are policy tools that bring in more resources to influence complex decisions that affect citizens in salient ways.

Aren't They Counterproductive to the Orderly and Rational Development of Policy?

No, without them and environmental organizations suing we would not have clean air standards. Most hard environmental decisions in the United States, with its 37-year-old baby of an environmental agency were hammered out in the adversarial world of the judiciary. Urban areas need to be brought into the maturing and more inclusive environmental agency of this century. British environmental programs all start with human habitation, as do most UNESCO and World Bank environmental programs. Global warming issues require the United States to deal with U.S. cities, in some way no matter who lives there. Treaty issues like Kyoto may arise.

Won't They Clog the Courts?

No. This is what courts do. When the EPA was formed it was designed to litigate, and it has litigated. Courts now have experience with aspects of environmental law like NEPA and so forth. Also, agencies are quickly self-righting when faced with a judicial order, so the cases against the U.S. agency would quickly decline.

ARE THE COURTS TOO POLITICAL WHEN DECIDING ENVIRONMENTAL DISPUTES?

Current trends in National Environmental Policy Act (NEPA) litigation in the federal courts show stark political polarization. Empirical research is being done to determine whether NEPA cases are being brought at the same rate as in the past; whether cases are being won more or less frequently; and whether the party affiliation of the presiding judge is correlated with the case outcome. The results are controversial and have opened up a battleground.

A 2001 report by Environmental Law Institute's Endangered Environmental Laws Program (*Judging NEPA: A "Hard Look" at Judicial Decision Making under the National Environmental Policy Act*, by Jay E. Austin, John M. Carter II, Bradley D. Klein, and Scott E. Schang) found that the rate of new NEPA litigation has increased significantly, in contrast to the general declining trend in NEPA litigation since 1974. This recent spike in litigation may be due to a number of factors, including an increase in environmentally significant actions by federal agencies, the above-mentioned U.S. Supreme Court decision, and a corresponding increase in challenges to those actions, or environmental parties' reported perception of the current administration as hostile to the values NEPA was designed to protect. Despite the increase in the number of new filings, the overall win rate for NEPA litigants has remained stable over 30 years of litigation.

There is a dramatic correlation between the outcome of NEPA cases and the political party affiliation of the presiding judge, using the party of the president who nominated that judge as a proxy. In 325 NEPA cases decided from January 21, 2001, to June 30, 2004,

- federal district court judges appointed by a Democratic president ruled in favor of environmental plaintiffs just under 60 percent of the time, while judges appointed by a Republican president ruled in their favor less than half as often, 28 percent of the time;
- district judges appointed by President George W. Bush have an even less favorable attitude toward environmentalists' NEPA suits, ruling in their favor only 17 percent of the time;
- when industry or pro-development interests sue under NEPA, the results are almost exactly reversed: judges appointed under a Democratic administration rule in favor of pro-development plaintiffs 14 percent of the time, while Republican-appointed judges rule in favor of such plaintiffs almost 60 percent of the time.
- a striking pattern can be found in the federal circuit courts, in which three-judge panels decide appeals from the district courts. Circuit court panels with a majority of judges appointed by a Democratic president (those with two or three such judges) ruled in favor of environmental plaintiffs 58 percent of the time. In contrast, Republican-majority panels ruled in favor of environmental plaintiffs 10 percent of the time, only one-sixth as often.

These results suggest that judges' political affiliations often have an impact on their decisions about NEPA claims. This is significant given the federal judiciary's central role in defining and enforcing NEPA's obligations. The results call into question whether NEPA is meeting its core purpose of providing a transparent environmental impact assessment process that generates information about proposed federal actions regardless of which administration proposes them, who objects to them, or who hears any dispute about them. Litigation of environmental issues under NEPA was written into the law. Nonetheless, the political polarization about the environment may have tainted the independence of the U.S. federal judiciary.

POTENTIAL FOR FUTURE CONTROVERSY

The foundation of modern environmental policy in the United States is a framework of court cases. The EPA was designed to handle litigation; many environmental laws have citizen-lawsuit provisions, and states may have separate citizen-suit provisions. This framework continues to this day. As long as environmental controversies continue, controversies about litigation as sound environmental policy making will continue.

See also Collaboration in Environmental Decision Making; Different Standards of Enforcement of Environmental Law; Environmental Audits and Environmental Audit Privileges; Environmental Justice; Environmental Mediation and Alternative Dispute Resolution

Web Resources

Florio, Kerry D. Attorneys' Fees in Environmental Citizen Suits: Should Prevailing Defendants Recover? Available at www.bc.edu/bc_org/avp/law/lwsch/journals/bcealr/27_4/03_FMS.htm. Accessed January 21, 2008.

States Environmental Resource Center. State Citizen Suits. Available at www.serconline.org/citizensuits/background.html. Accessed January 21, 2008.

Further Reading: Brooks, Richard O. 1996. *Green Justice: The Environment and the Courts.* Boulder, CO: Westview Press; Chen, Jim, and Jonathan H. Adler, eds. 2003. *The Jurisdynamics of Environmental Protection: Change and the Pragmatic Voice in Environmental Law.* Washington, DC: Environmental Law Institute; Collin, Robert W. 2006. *The US Environmental Protection Agency: Cleaning Up America's Act.* Westport, CT: Greenwood; Lazarus, Richard J. 2004. *Making of Environmental Law.* Chicago: University of Chicago Press; Schoenbrod, David. 2005. *Saving Our Environment from Washington: How Congress Grabs Power, Shirks Responsibility, and Short Changes the People.* New Haven, CT: Yale University Press; Wilkinson, David. 2002. *Environment and Law.* New York: Routledge.

LOGGING

Logging trees has severe environmental impacts. Attempts at sustainable logging are considered monoculture production, not ecosystem preservation. The U.S. government permits its land to be harvested for timber, with the profit going to private corporations. Permits for cutting trees, salvage logging, and fire-reduction cuts are all controversial.

TREES AND LOGGING: HISTORIC DUAL PURPOSE, MODERN-DAY RAW MATERIAL

Historically, clearing land of trees was necessary for human settlement. The land itself was necessary for growing crops. The trees provided cover for wildlife, which most pioneers tried to keep away from the house and livestock. Today, clearing trees by centuries-old slash-and-burn techniques continues. One big difference is that today, with modern technology and monocultural pesticide-supported growing techniques, the scale of the deforestation is much larger and faster. The deforestation controversy combines with logging controversies.

Today, the land-clearing function of logging is a smaller, though still controversial, part of logging. Logging now treats trees as raw materials for the production of other products such as wood, paper, and others.

ENVIRONMENTAL IMPACTS OF LOGGING

Logging has environmental impacts in both the short and long term. It affects many components of the ecosystem. Trees are essential components in many ecosystems, retaining water, providing shade over land and water, as well as providing food and shelter to wildlife.

There are many types of trees, although only some are logged. In many ecosystems trees fill unique ecological niches. One example is the Port Orfid cedar growing on Oregon's coast. Having adapted to the wet, windy conditions of its ecosystem, it flourished. When it was plentiful, it was logged. In the 1980s a deadly virus attacked the tree. Port Orfid cedars grow and communicate through their roots, and the virus moves from tree to tree, wiping out whole groves. State biologists scrambled for preventative solutions, and environmentalists became very concerned. The U.S. Forest Service permitted logging in and near a wilderness area with some of the last stands of Port Orfid cedar. The loggers transmitted the virus to these trees although they were not logging the Port Orfid cedars. Intense environmental litigation ensued. Among other issues the court ruled that the loggers were to wash off their logging vehicles when entering and leaving the wilderness area. They do not do this, and this battleground simmers. This is an example of the environmental impact of logging on other parts of the ecosystem, including endangered species like the spotted owl. Making roads into natural or manmade wilderness areas creates ecosystem-threatening risks.

LOGGING AND SOIL EROSION

A big battleground in logging is whether and how much soil erosion it causes. Soil erosion has a bad environmental impact in most places because it lowers the ecological productivity of a region. Plants and animals depend on soil or on the plants and animals the soil feeds. When roads are built and heavy loads of logs are moved on them, the impact on the stability of the soil on any type of slope is enormous. Since many trees are in mountainous areas, logging roads are often built along the sides of valleys or canyons. Some states have more miles of logging roads than paved roads. At the bottom of these valleys and canyons are usually creeks, streams, and rivers. When the soil becomes unstable it can cause landslides. These landslides can occur months or years after the logging operation ceases because the trees are no longer there to hold the water and soil in place. Forests provide a buffer that filters water and holds soil in place. They sustain water and soil resources by recycling nutrients. In watersheds where forests are environmentally impacted by logging, the soil erosion results in siltation of the waterways, which, combined with the loss of shade from trees, increases the temperature of the water, affecting the ability of aquatic life to exist. It can also affect the ability of humans to use the water downstream. This affects agribusiness users, cities, and natural systems. One impact that overlogging caused in Salem, Oregon, in the late 1990s is the closure of the public water supply. Clearcutting trees on steep slopes had caused siltation of the water in the watershed. This particular battleground developed in the state legislature, which is a common battleground for logging issues.

LOGGING AND CLIMATE IMPACTS

Local changes in precipitation are direct and immediate when the forest cover is removed. Trees and forests engage in a natural process of transpiration that modifies water flow, softening the impact of highs and lows in water volume. Changes in transpiration increase both runoff and erosion.

LOSS OF BIODIVERSITY

Logging Often Destroys Natural Habitats for Long Periods of Time

Environmentalists claim that logging destroys natural habitats, resulting in the loss of biodiversity. They fear that that this could lead to the extinction of species, such as the spotted owl. The logging industry has developed some replanting and selective-resulting harvesting techniques to minimize habitat degradation and soil erosion. Some logging corporations say they try to keep a protected strip of land next to the waterways, called riparian protection. They limit the amount of trees they take from steep slopes to prevent soil erosion. They buy land and plant their own trees and engage in logging there, as opposed to in national parks, wilderness areas, and other federally owned land. Logging corporations also prepare environmental impact statements when trying to cut trees on federal land. Each one of these environmental mitigation techniques and best practices is a battleground contested by environmentalists who say logging corporations do minimal mitigation, create far more impact than they self-report, and take trees illegally from federally owned lands. Many rural western communities are dependent on logging as the only source of income. Since the spotted owl litigation, about 60 logging mills closed down in Oregon alone, leaving some communities with no economic base.

Sustainable Logging

Sustainable forestry epitomized the ideal that the land would never be depleted of its trees. As early as 1936, Weyerhaeuser practiced sustained-yield logging, a practice it continues in the Willamette Valley of Oregon. By logging the forest of its timber and replacing a variety of tree species with a single commercial species timber companies create a continuing supply of timber. There may also be less environmental impacts than when logging virgin or secondary-growth forests. Environmentalists accuse logging companies of planting monocultures, with additional environmental impacts because of pesticide use and susceptibility to disease. They point out that simply planting your own trees does not necessarily lessen the environmental impacts, especially to the soil. Many small property owners in some of the timber states have stands of timber they have planted to harvest as a future investment. Some of these stands can contain unique and irreplaceable old-growth forest. When environmentalists pursue state legislative changes that alter expectations for use of this timber, private property activists get involved and open up another battleground in the logging controversy. When loggers cut old-growth forest, environmental activists increase protest activities such as tree sitting. When timber corporations begin sustainability programs, it is in the context of this bitterly entrenched, litigation-heavy controversy.

ROADLESS AREAS AND LOGGING

Many environmental laws protect federal lands as well as control multiple uses on them. Roads are a major concern because they erode forever-wild status in wilderness areas and have large environmental impacts. Other recreational users, such as all-terrain vehicles, mountain bikers, and packhorses, all begin to use the road. In this way, building roads so that corporations can mine, log, or graze animals opens up that area to a longer term, much more pervasive environmental impact. Roadless areas cover approximately 50 to 60 million acres, or about 25 to 30 percent of all land in the national forests; another 35 million acres are congressionally designated wilderness. The rest of the national forests contain 380,000 miles of roads. The vast majority of these rough-cut roads are built to provide access for logging and, to a much lesser extent, mining. Public opposition to subsidized logging of public lands by building roads, bridges, and aqueducts in parks and wilderness areas resulted in budget cuts for the Forest Service. Watershed protection and restoration, or timber production, became the outlines for this battleground. Environmentalists have long accused the Forest Service of simply managing the profits for industry, not protecting the environment. The arena of public funding of roads in these areas is a narrow but well-defined battleground.

There is substantial public and political support for permanent protection of the roadless areas. Historically, many conceived of national parks and wilderness areas as being without any roads. With the increased reliance on the car, roads were developed everywhere. Roads greatly impact pristine natural areas, so much so that the need to protect roadless areas is well known. A driving need in environmental policy is to establish a long-term policy that protects all roadless national forest areas from road building, logging, mining, grazing, and other environmentally degrading activities.

In 2004 the Bush administration proposed a new plan to open up national forests to more logging. It focuses on road building. Under Bush's plan, governors would have to petition the federal government to block roads needed for logging in remote areas of national forests. It covers about 58 million of the 191 million acres of national forest nationwide. Since 2002, the George W. Bush administration has been studying the roadless rule, which blocks road construction in nearly one-third of national forests to prevent logging and other commercial activity. The timber industry supports the proposal, maintaining that these decisions are far better made by the local community and state than through federal policy.

But New Mexico's Governor Bill Richardson, said the Forest Service was walking away from environmental protection. Richardson said he would ask that all 1.1 million acres of roadless land in his state remain protected and planned to urge other western governors to do the same.

Environmental groups were also critical. "This is a roadblock to roadless protection. To take it down to the state level like this really undermines having a national forest system," added the Wilderness Society. National Environmental Trust called the proposal "the biggest single giveaway to the timber industry in the history of the national forests." Without a national policy against road construction, environmentalists maintained, forest management would revert to individual forest plans that in many cases allowed roads and other development on most of the 58 million acres that are now protected by the roadless rule. Environmentalists say it is unlikely that governors in pro-logging states, such as Idaho, Wyoming, Montana, and Utah, will seek to keep the roadless rule in effect. Several Republican governors in the West have strongly criticized the rule, calling it "an unnecessary restriction that has locked up millions of acres from logging and other economic development."

LOGGING, ENDANGERED SPECIES, AND SNAGS

The environmental impacts of logging often threaten endangered species such as the spotted owl, which spawned a decade of political and legal controversy in Oregon. In San Francisco, a federal district court granted an environmentalist's request for a halt to a logging project that would cut thousands of very large dead trees (called snags) from a burn area that includes critical habitat for the northern spotted owl. This area of snags is part of an old-growth forest reserve in the Six Rivers National Forest in northern California. The controversy swirls around the Sims Fire Salvage timber sale.

According to federal government reports, the Sims Fire Salvage timber sale would cut an estimated 6.1 million board feet from 169 acres, an average of more than 36,000 board feet per acre. The average tree to be cut is nearly three feet in diameter and many are larger. The Six Rivers National Forest has designated Sierra Pacific Industries the high bidder for the Sims sale but has not yet awarded a contract to the company. Of the 169 acres that would be cut, 124 are part of a late successional reserve, which environmentalists contend was to be set aside to provide old-growth habitat.

The sale area also includes 57 acres of designated critical habitat for the northern spotted owl, critical habitat for the marbled murrelet and critical habitat for the coho salmon. All three species are listed as threatened under the federal Endangered Species Act. Environmentalists charge that the federal agency is emphasizing logging projects at the expense of its responsibility to protect fish and wildlife habitat in the public forests.

Dead trees provide important wildlife habitat, and may serve other functions in ecological processes. Snags hold fragile soils in place and provide critical wildlife habitat. Environmentalists have scientists and law on their side. This court case, and others like it, will continue to define the ragged edges of U.S. environmental policy around logging. This is also one of the primary battlegrounds of this environmentally controversial issue.

POTENTIAL FOR FUTURE CONTROVERSY

With the fate of the roadless areas in doubt and industry scientists, environmentalists, and federal agencies intensely studying these issues, the logging controversy will continue. As timber supplies dwindle, governmental agencies may need to take private property to control ecosystem impacts from logging. Some logging communities live and die based on local mill operations. Other communities are concerned that their watershed could be contaminated from runoff or spraying from logging. Logging is also an international environmental issue, and global developments around logging may affect the U.S. controversy.

See also Environmental Impact Statements: United States; Federal Environmental Land Use; Fire; State Environmental Land Use; "Takings" of Private Property under the U.S. Constitution; Watershed Protection and Soil Conservation

Web Resources

Andre, Claire, and Manuel Velasquez. Ethics and the Spotted Owl Controversy. Available at www.scu.edu/ethics/publications/iie/v4n1/. Accessed January 21, 2008.

Knickerbocker, Brad. With Oregon Timber Sale, Controversy Flares. *Christian Science Monitor,* June 14, 2006. Available at www.csmonitor.com/2006/0614/p02s01-usgn.html. Accessed January 21, 2008.

Roadless Area Conservation. Available at www.roadless.fs.fed.us/. Accessed January 21, 2008.

Further Reading: Alverson, William Surprison, Walter Kuhlmann, and Donald M. Waller. 1994. *Wild Forests: Conservation Biology and Public Policy.* Washington, DC: Island Press; Goble, Dale, and Paul W. Hirt. 1999. *Northwest Lands, Northwest Peoples: Readings in Environmental History.* Seattle: University of Washington Press; Keiter, Robert B. 2003. *Keeping Faith with Nature: Ecosystems, Democracy, and America's Public Lands.* New Haven, CT: Yale University Press; Knott, Catherine Henshaw. 1998. *Living with the Adirondack Forest: Local Perspectives on Land Use Conflicts.* Ithaca, NY: Cornell University Press; Marchak, M. Patricia. 1995. *Logging the Globe.* Kingston, Ontario: McGill-Queen's Press; Rajala, Richard A. 1998. *Clearcutting the Pacific Rain Forest: Production, Science, and Regulation.* Vancouver: University of British Columbia Press.

LOW-LEVEL RADIOACTIVE WASTE

The transportation, shipment, storage, and disposal of low-level radioactive materials may cause people to come into contact with them. States and communities actively resist these types of exposures. Industry maintains that low levels of exposure are not risky.

WHERE DOES RADIOACTIVE WASTE COME FROM?

Nuclear power and research generate low-level radioactive waste (LLRW). Twenty percent of U.S. energy comes from nuclear energy, and other countries are increasingly dependent on nuclear energy. As environmental controversies such as global warming and climate change drive a search for more alternate energy sources, demand for nuclear energy is likely to increase. Low-level radioactive waste production may also increase. Radioactive materials can present risk to all life-forms for many years.

The term *low level* does not mean that LLRW is relatively harmless, nor does it not mean *low hazard* to human health. There has been little consideration of the synergistic relationships of radiation and other environmental contaminants upon an individual recipient. The uncertainty of the exposure to this dangerous and increasing material increases concern and controversy.

WHO WILL TAKE THE WORST WASTE?
THE CASE OF SKULL VALLEY, UTAH

Utah is a dry western state in which much of the land is owned by the federal government. There are also many Indian reservations there. The Goshutes people have inhabited that part of the land for many years. They were there before the Mormons, the Mexicans, and the Spaniards. At their peak the Goshutes numbered about 20,000. Today there are fewer than 500 Goshutes, of which 124 belong to the Skull Valley Band (http://skullvalleygoshutes.org).

The Goshutes were recently issued a license from the Nuclear Regulatory Commission to accept radioactive and other hazardous wastes.

Today, the Skull Valley Goshute Reservation comprises approximately 18,000 acres. South of Skull Valley on traditional Goshute territory is now the Dugway Proving Grounds. This is where the U.S. government developed and tested chemical and biological weapons. In 1968 chemical agents leaked into the environment from Dugway, and approximately 6,000 sheep and other animals died. At least 1,600 of those contaminated sheep were buried on the Goshutes' reservation by the U.S. government.

Dumping Grounds

Skull Valley is the middle of a toxic donut. It is surrounded by dangerous and toxic land uses. East of Skull Valley in the area known as Rush Valley, there is a nerve gas storage facility for the federal government. The world's largest nerve gas incinerator has recently been built to destroy thousands of tons of these deadly chemicals. To the south of Skull Valley is the Intermountain Power Project, which provides coal-fired electrical power primarily for California. It also produces air pollution that affects the Skull Valley Reservation. Northwest of Skull Valley is the Envirocare Low-Level Radioactive Disposal Site. This corporation buries radioactive waste for the United States. Close by are also two hazardous waste incinerators and one hazardous waste landfill. North of the reservation is the Magnesium Corporation plant, a large magnesium-production plant. This is the most polluting plant of its kind in the United States. Chlorine gas releases from Magnesium Corporation also affect the Skull Valley Reservation. In the siting of all these facilities on the aboriginal territory of the Goshutes, the Skull Valley tribal government and people were never once consulted (http://skullvalleygoshutes.org).

Current Controversy

Native people are well aware of their environmental conditions and, like most communities, seek economic development. Many reservations have high unemployment rates and lack economic opportunity. The band considered a variety of economic ventures, including the storage of spent nuclear fuel. The Skull Valley Band of Goshutes leased land to a private group of electrical utilities for the temporary storage of 40,000 metric tons of spent nuclear fuel.

The proposition to store spent nuclear fuel on Native American reservations began with a controversial proposal in 1990 by the Office of the Nuclear Waste Negotiator to communities nationally. The federal government sought a voluntary candidate site to consider the temporary storage of spent fuel as provided by the Nuclear Waste Policy Act of 1982 and the amendments of 1987. The storage program was known as monitored retrievable storage or MRS. When the executive committee of the Goshute Band became aware of this proposal, they submitted a grant application and began gathering data. The $100,000 grant was awarded in 1992 for the band to investigate the benefits and impacts of siting an MRS at Skull Valley. From 1992 until 1995, the leaders of the band carefully studied the facts. They made site visits to examine firsthand all aspects of storage of spent nuclear fuel under the MRS program.

Why Can the Goshutes Not Voluntarily Take Deadly Waste for Disposal?

The opposition to this project has been intense and politically charged. Some of the foremost nuclear scientists have decided to intervene in the proceedings before the Nuclear Regulatory Commission (NRC). They represent Scientists for Secure Waste Storage (SSWS). These scientists seek to provide objective scientific evidence in the NRC proceedings. Their intervention does not constitute an endorsement or opposition to this project. SSWS is represented by the Atlantic Legal Foundation, Inc., a nonprofit law firm that brings scientific clarity to important national cases. They are concerned that the Private Fuel Storage (PFS) facility could cause radioactive contamination of the groundwater. The Goshutes have been saying that their storage casks will be raised above the anticipated floodwater, and floodwater and rainwater will not enter the storage site. Regular monitoring can help prevent this before any groundwater is contaminated, before it leaves the reservation, and before it reaches other people.

The Goshutes have agreed to accept 10,000 pounds of hazardous and toxic waste from Kobe, Japan, as a first shipment. The site will eventually hold 700 tons of low-level radioactive wastes, attracting radioactive waste streams from all around the world. It will have to be shipped, trucked, delivered, and stored all along the way, presenting long-term risk if spilled or leaked. Environmentalists are very concerned by a process that was fast and noninclusive, and by a result that sets a dangerous precedent. Some are questioning whether this is an abuse of Native American sovereignty by industry. Others claim that it is an environmental justice issue because the Goshutes were already disproportionately affected by other risky and environmentally degrading uses.

Battleground: Risk to Public Health and Environmental Impacts

There are more than 2,500,000 shipments of low-level radioactive waste per year in the United States by highway, railroad, and air. During the 10-year period of 1971–1981, there were 108 accidents involving the transport of that waste. Four of these accidents involved spent fuel casks.

The waste site is very large. Some waste may come via ports to roads and railroads. Many communities express concern about these materials around them. Some cities designate certain routes for vehicles carrying hazardous materials. Utah's Senator Orrin Hatch has publicly expressed concern that much of this radioactive waste is coming from foreign countries. Many elected officials and state agencies in Utah now claim they were unaware of this fact when approvals were sought. This battleground currently involves the Utah Attorney General and will involve many courts and public hearings.

In the United States, regulation of radioactive waste is divided into two major categories, high-level radioactive waste and low-level radioactive waste. There are two other very narrowly defined special categories, which will be discussed further on. Low-level waste is generally defined as any radioactive waste that does not belong in any of the other categories. As a result, low-level waste is a very broad catchall category comprising many different types of waste and a wide range of concentrations of radioactive materials. Low-level waste is further divided into four classes of waste with specific regulations for the disposal of each class.

DEFINITION OF LOW-LEVEL RADIOACTIVE WASTE

Low-level radioactive waste is defined by law and enforced by federal agencies in the US. This itself can be a matter of controversy because the lowest level of radioactive waste is often the least regulated. The regulatory thresholds for public safety from exposure to low level radioactive waste do not consider issues of cumulative impacts or human differences in vulnerability. According to law, low-level radioactive waste is defined as any radioactive waste that does not belong in one of three categories. Again, according to law, those other three categories are

1. high-level waste (spent nuclear fuel or the highly radioactive waste produced if spent fuel is reprocessed),
2. uranium milling residues, and
3. waste with greater than specified quantities of elements heavier than uranium.

Spent nuclear fuel means fuel from nuclear power plants. Spent fuel sometimes contains some reusable material that may be recovered in a process called reprocessing. Basically, everything left over after the reusable material has been recovered is classified as high-level radioactive waste. The United States is not reprocessing spent nuclear fuel at this time. This too can be a controversial flashpoint because some claim that more efficient reprocessing can lessen environmental impacts and reduce human risks from exposure.

Low-level radioactivity often exists in mine tailings. Uranium milling leaves behind rock and soil that remain after uranium has been removed from the ore. These milling residues are also known as tailings. Radioactive waste that contains more than a specified concentration of elements heavier than uranium is called transuranic. It is not classified as low-level radioactive waste. All other radioactive waste is low-level radioactive waste. In this manner, the category of low-level radioactive waste is a catch all term.

Low-level radioactive waste is generated by commercial operations, Department of Defense activities and U.S. Department of Energy operations. The federal government is responsible for disposal of low-level waste generated by the Department of Energy. The states are responsible for disposal of low-level waste from (1) commercial operations such as utilities, industries, hospitals, and research institutions, and (2) government facilities such as veterans' hospitals and nonweapons-related government facilities.

CLASSES OF COMMERCIAL LOW-LEVEL RADIOACTIVE WASTE

Commercial low-level radioactive waste is further divided into three classes, again by law. Three classes of commercial low-level radioactive waste are defined in the Code of Federal Regulations, Title 10, Part 61. Those classes are Class A, Class B, and Class C. These regulations list the limits on concentrations of specific radioactive materials allowed in each low-level waste class. Radioactive waste not meeting the criteria for these classes falls into a fourth class, known as Greater Than Class C. According to the law, class A low-level radioactive waste contains the lowest concentration of radioactive materials, and most of those materials have half-lives of less than five years. Class B contains the next lowest concentration of radioactive materials, and it contains a higher proportion of materials with longer half-lives. Class C low-level waste has the highest concentration of radioactive material allowed to be buried in a low-level waste disposal facility. The concentration of radioactive materials in Greater Than Class C exceeds the limits for Class C waste. This is the beginning of another environmental controversy. All Greater Than Class C waste is the responsibility of the federal government and must be disposed of in a geologic repository such as the high-level waste repositories planned for Nevada and Utah.

AMOUNT OF LOW-LEVEL RADIOACTIVE WASTE PRODUCED

The amount of low-level radioactive waste produced is usually described in one of two ways. One way is by volume of the waste (in cubic feet or cubic meters). The other way is to give the activity. Activity is the rate at which radiation is given off by the material in the waste; it is measured in curies. These methods of measurement are prescribed by law.

The volume and activity of low-level radioactive waste generated and shipped in the United States each year can vary significantly. Nuclear power plants, industries, medical facilities, research institutions, defense agencies, military installations, and universities generate low-level radioactive waste. This low-level radioactive waste must be shipped to a specially designed facility for permanent disposal.

Government regulations define the three classes of low-level radioactive waste that may be placed in a low-level waste disposal facility. The definitions of these classes are based on the specific radioactive isotopes present, their concentrations, and their half-lives. Any low-level waste with a concentration of radioactive material greater than that specified for these three classes is a federal responsibility. This means that it must be disposed of in a deep geologic repository like Yucca Mountain in Nevada, or Skull Valley, Utah. Again, this is where another environmental controversy begins.

METHODS OF DISPOSAL

Historically, two methods of low-level radioactive waste disposal have been used: (1) near-surface land disposal and (2) ocean disposal. Near-surface land disposal involves confining low-level waste either at or below the earth's surface. Ocean disposal, employed by the Atomic Energy Commission prior to 1970, involved depositing waste containers on the ocean floor. The Atomic Energy Commission was a federal agency that helped the young nuclear energy industry regulate itself. This method is no longer used. Other disposal options such as mine disposal, deep well injection, and beneath-seabed disposal are used or being considered around the world but are not currently used or under consideration in the United States. All these methods are very controversial.

The technology of radioactive waste disposal remains experimental. Stability of waste forms remains controversial. A June 1996 report on microbial degradation of cement issued by the Nuclear Regulatory Commission states:

> Testing conducted with the developed biodegradation test has convincingly demonstrated that cement-solidified LLW waste forms can be attacked and degraded by the action of ubiquitous microorganisms that are present at LLW disposal sites. It was shown that during the degradation process, large percentages of those elements composing the cement matrix of waste forms were removed. In addition, it was conclusively shown that the ability of cement-based waste forms to retain or retard the loss of encapsulated radionuclides *was compromised* [emphasis added] due to the action of microorganisms.

While this form of research does show some potential, it does not prove that low-level radioactive waste is biodegradable.

POTENTIAL FOR FUTURE CONTROVERSY

Concern about low-level radioactive waste among communities and environmental advocates shows no sign of decreasing. State and city governments are engaged in resisting these sites in many places. They can do so by fighting site selection and limiting transportation options through their communities. Other communities have chosen to accept this waste, such as the Skull Valley Band of the Goshutes. Communities that oppose low-level radioactive waste in their midst generally have a local battleground, with state and local legislators often involved in local zoning and land-use boards. Communities that seek low-level radioactive waste have a federal battleground in both legislative and judicial forums. Nearby communities, the host state, environmentalists, sustainability advocates, and others strongly oppose low-level radioactive waste anywhere.

See also Cumulative Emissions, Impacts, and Risks; Ecosystem Risk Assessment; Environmental Justice; Human Health Risk Assessment; Precautionary Principle; Toxics Release Inventory

Web Resources

Disposal of Radioactive Waste in Goshutes Land. Available at skullvalleygoshutes.org. Accessed January 21, 2008.
U.S. Nuclear Regulatory Commission. Low-Level Waste. Available at www.nrc.gov/waste/low-level-waste.html. Accessed January 21, 2008.

Further Reading: Berlin, Robert E., and Catherine C. Stanton. 1989. *Radioactive Waste Management.* New York: John Wiley and Sons; Burns, Michael E., ed. 1988. *Low-Level Radioactive Waste Regulation: Science, Politics and Fear.* Chelsea, MI: Lewis Publishers; Cooper, John R., Keith Randle, and Ranjeet S. Sokhi. 2003. *Radioactive Releases in the Environment: Impact and Assessment.* New York: John Wiley and Sons; Gershey, Edward L., et al. 1990. *Low-Level Radioactive Waste from Cradle to Grave.* New York: Van Nostrand Reinhold; Weber, Isabelle P., and Susan D. Wiltshire. 1985. *The Nuclear Waste Primer: A Handbook for Citizens.* New York: The League of Women Voters Education Fund, Nick Lyons Books.

M

MINING OF NATURAL RESOURCES

Mining is the process of extracting ores and other substances from the earth. It can have enormous and irreparable environmental impacts, especially with technological improvements in the industry. These impacts affect national parks, indigenous peoples, and nearby communities. In the United States, many large mines are on land leased from the government to private corporations. Some communities are dependent on the mining industry, such as those around some coal mines.

BACKGROUND

The 1872 Mining Law allowed the mining of valuable minerals on federal land with minimal payments to the U.S. government. Its purpose was to encourage westward expansion of European settlement. Some of the oldest roads in the west are old mining roads. Mining for gold, silver, and other minerals was extremely dangerous work in the late 1800s. The mines were very warm, collapsed frequently, were subject to fires and floods, and were filled with toxic gases. Long-term leases, low-cost sales, and other arrangements allowed mining interests to develop a basic natural resource in the west. While doing so, some of the basic road infrastructure was developed. The profits for these government-protected risks is one of the battlegrounds. Critics of the 1872 Mining Law contend that the profits generated by mining federal lands are very large and no longer need any government subsidization. Environmental concerns about roads generally, and about the increasing scale and environmental impacts of mining and loss of habitat, enter the battleground.

SPECULATION ON FEDERAL LANDS: ENVIRONMENTAL IMPACTS AND CONTROVERSIES

Mining is restricted by local land-use regulations, state environmental laws, and federal environmental rules and regulations. Many more restrictions are imposed on the timing of mining activities on federal land. Generally, environmentalists would like to see more mitigation and cleanup of environmental degradation. Many environmentalists would like to see absolutely no mining in areas where there are endangered species. In terms of land speculation with federal mineral rights leases, the issue is how long the lease can be held without mining. This makes it difficult for things such as conservation easements, or any private property owner wishing to simply not develop his or her mineral right. It is a use-it-or-lose-it proposition that works to increase mining and the environmental impacts of mining. Diligence requirements in the leases limit how long a lease can be held without any development and how long it can be held after production is shut down. Moreover, regular expenditures are required by the terms of the lease. Environmentalists and others maintain that those restrictions are not

CHILD LABOR IN MINING

According to the second Global Report on Child Labor prepared by the International Labor Organization (ILO) the end of child labor could occur in our lifetime. Prepared under the ILO's Declaration on Fundamental Principles and Rights and Work and the International Programme on the Elimination of Child Labour, the new report, titled "The End of Child Labor—Within Reach," says there is a reduction in child labor in many parts of the world. If current trends continue, they claim, child labor in its worst forms may be eliminated within the next decade. Some of child labor's worst forms occur in mining.

The report indicates that the number of child laborers globally has fallen by 11 percent over the last four years, that is, 28 million fewer than in 2002. The sharpest decrease is in the area of hazardous work by children. There has been a 26 percent reduction overall, and 33 percent fewer children between the ages of 5 and 14 who are endangering their lives in hazardous work. The numbers can be hard to obtain for this issue because employers often hide children laborers.

Children who work in the mining sector are at particular risk, the report noted. Mining in many parts of the world is still low tech. It is very risky in the short and long term. The ILO estimates that some one million children work in small-scale mining and quarrying around the globe. In an area as complex as child labor and mining it is likely that these numbers could be much worse in regions where counting does not occur. Cultural differences about "childhood" and "work" abound, as do gender differences within and between cultures. Children are small and energetic. They can reach and go places in mining an adult could not reach. They can be subject to life-threatening situations because of both work and poverty. Other cultures may teach the value of work by assigning chores of increasing responsibility. In mining, technology has played a major role in decreasing worker exposure and increasing worker safety. It also decreased the need for miners.

rigorous enough to constrain development. Others think that restrictions are a good idea but that existing restrictions are more than adequate.

SHARING THE WEALTH: WHAT TO DO WITH MINING REVENUES

Mining fees are distributed primarily to residents of sparsely populated western states because Congress allocates half of gross mining receipts to the state in which the mining occurs. That is one reason many state environmental agencies and communities in these states support the mining industry. Many environmentalists would like to use some of that money to restore the ecology to its premining ecological condition.

SUSTAINABILITY AND MINING

Many question whether mining can be described as sustainable. Of all the earth and ore disturbed for metals extraction, only a small amount is actual ore. For example, in 1995 the gold industry moved and processed 72.5 million tons of rock to extract 7,235 tons of gold. The rest, 99 percent, was left as waste. Mine tailings can be hazardous and build up quickly in the host community. Cleanup of radioactive uranium tailings is an environmental battleground. Some Native American environmental justice issues revolve around the cleanup of low-level radioactive waste, often piles of mine tailings. These aspects of this controversy form its battlefield.

An early environmental battleground is whether it is acceptable to mine. For example, in some instances, even an operation with state-of-the-art environmental design should simply not be built because it is planned for a location that is not appropriate for mining. Environmental critics claim that mining companies want to engage sustainability only in terms of how to mine, not whether to mine.

COMMUNITY CONCERN

While nations and multinational corporations profit from mining operations around the world, local communities face the resulting environmental impacts. Mining communities have begun to exercise their right to prior informed consent to mining operations.

The concept of prior informed consent is the right of a community to be informed about mining operations on a full and timely basis. It allows a community to approve an operation prior to commencement. This includes participation in setting the terms and conditions and addressing the economic, social, and environmental impacts of all phases of mining and postmining operations. Some environmentalists oppose this type of community rule because communities' short-term economic interests may outweigh long-term environmental conditions. They wonder how well all the terms and conditions in the prior informed consent would really be enforced.

POTENTIAL FOR FUTURE CONTROVERSY

Communities, environmentalists, mining companies and their employees, and government all decry the environmental impacts of mining. Yet consumer demand for products made from mined materials and a rapid increase in technology allow the scope and scale of mining to increase. This will increase environmental impacts and also controversy.

See also Cumulative Emissions, Impacts, and Risks; Environmental Impact Statements: International; Environmental Impact Statements: Tribal; Environmental Impact Statements: United States; Sustainability; Water Pollution

Web Resources

The 3rd World View. Canary in the Coal Mine: What the Deaths in Phulbari Mean to Bangladesh. August 29, 2006. Available at rezwanul.blogspot.com/2006/08/canary-in-coal-mine-what-deaths-in.html. Accessed January 21, 2008.

U.S. Department of the Interior, Office of Surface Mining, Environmental Assessments. Available at www.osmre.gov/pdf/streambufferea.pdf. Accessed January 21, 2008.

Further Reading: Crowder, Ad'aele A., Earle A. Ripley, and Robert E. Redmann. 1996. *Environmental Effects of Mining.* Singapore: CRC Press; Hartman, Howard L., and Jan M. Mutmansky. 2002. *Introductory Mining Engineering.* New York: John Wiley and Sons; Hester, Ronald Ernest. 1994. *Mining and Its Environmental Impact.* London: Royal Society of Chemistry.

MOUNTAIN RESCUES

When hikers, climbers, and boaters need emergency assistance, the rescue is often performed by local government. With powerful cell phones, the ability to call in a rescue request is increased. Rescuers are often put at risk looking for lost and injured people. A controversial issue is whether to make those rescued pay for their rescue.

Mountain rescue refers to search-and-rescue activities that occur in mountains, although the term is sometimes also applied to search and rescue in other wilderness environments. The increase in cell phones, geographic positioning systems, and other technology has allowed recreational mountain hiking, climbing, and skiing to increase. Many have always assumed that if you could communicate with help, you could get help. With modern technology this may not be the case, and it forms the contours of the battleground. The difficult and remote nature of the terrain in which mountain rescue often occurs makes it extremely risky for rescuers. Often rescues occur during bad weather conditions, increasing risks to all involved.

Mountain rescue services may be professional or volunteer. Professional services are more likely to exist in places with a high demand such as the Swiss Alps, national and provincial parks with mountain terrain, and many ski resorts. The labor-intensive and sporadic nature of mountain rescues, along with the specific techniques and local knowledge required for some environments, means that mountain rescue is often by voluntary teams. These are frequently

RACISM AND RESCUES?

What if you are lost in the mountains and your car gets stuck in a two-day snowstorm? You hope to be rescued. The usual advice is to stay with the vehicle until help arrives. But what if it does not? How long do you wait? Two contrasting mountain rescues have unleashed a storm of controversy.

In December 2006, three mountaineers from Texas and New York attempted a very dangerous ascent of Mt. Hood in a blizzard, outfitted for a quick ascent. They were experienced, strong, white males in the prime of their lives. When the 130-mile-per-hour winds tried to blast them off the mountain, they built snow caves to survive. The local sheriff, the National Guard, other military resources, the Oregon State police, and many volunteers surrounded the mountain until the weather broke four days later. Using heat-seeking devices and unmanned small aircraft they tried to find the men. They did not. Using rescuers on the ground, dropped by helicopter, they eventually found the snow caves and the body of one climber. The bodies of the other two have not been recovered. This was an expensive search. The families flew to Timberline Lodge, the nearest location, and worked closely every day with the teams of rescuers. No other people were lost in this search, a major accomplishment given the weather conditions, number of people involved, and intense media scrutiny.

This was preceded by another mountain-rescue search in Oregon. James and Kati Kim and their two young daughters got lost the night of November 25, 2006, trying to drive a backcountry route known as Bear Camp Road through the Siskiyou National Forest during a snowstorm. They were headed for a coastal hotel where they had reservations. They were not reported missing for four days, and the initial search stretched more than 300 miles between Portland and the coast. After being stranded a week on a remote logging road that branches off Bear Camp Road, James Kim hiked for help on December 2. He left the road to follow a creek, where he was found dead of exposure four days later.

Two days after he left, his wife and two young daughters were found by a local private helicopter pilot who was following a hunch and was not involved in the formal search. The Kims' family hired its own helicopters to join the search.

Two Edge Wireless engineers parsed cell-phone records and found that a text message to the Kims had bounced off a cell-phone tower near Glendale and been received somewhere to the west of the tower. They notified authorities the night of December 2. Oregon sheriffs did not respond to this very important clue until more than 24 hours later, and this is one aspect of the current debate.

The *Oregonian* newspaper suggested officials in Josephine County did not make effective use of tips about what road the Kims might have taken. A critical point in this critique is finding out when the search came under the control of one county. In the early stages everybody thought it was in a different county.

The San Francisco father and CNET editor died of exposure to the wet and cold, bravely trying to get help for his family. The *Oregonian* reported the emergency services coordinator for the area felt overwhelmed and waited two days to call for National Guard helicopters. The paper also said the coordinator's phone call was ignored by a supervisor who was at home watching football on his day off.

The key to finding them, police said, was a "ping" from one of the family's cell phones that helped narrow down their location. James Kim died of exposure with hypothermia while trying to walk for help, according to the Oregon state medical examiner. Helicopter search crews hired by James Kim's father found the body in a ravine in the Oregon wilderness. His body was found at the foot of the Big Windy Creek drainage, a half-mile from the Rogue River, where ground crews and helicopters had been searching for days. It was less than a mile from where his family's car got stuck in the snow.

James Kim, 35, was a senior editor for San Francisco–based technology media company CNET Networks Inc. He was an Asian American married to a white woman with small children. He broke down in a part of Oregon noted not only for scenic beauty but also the possible presence of controversial groups such as the Ku Klux Klan. He and his family were found by privately hired rescuers paid for by the family. If the local sheriff had reacted to the cell-phone pings with the same alacrity as the sheriffs did on Mt. Hood, many feel that Mr. Kim could be alive. Others are incredulous that they could not find a family of four on a paved road in a location where others have required rescue, especially since they burned their tires to attract attention.

It is very difficult to infer intentional acts of racism from the failure of a governmental official, such as a sheriff, to act. Yet, these two searches are starkly different. The Kim rescue is the subject of several continuing investigations by different agencies. No court cases have been filed to date, but there is still time to do so.

---

made up of local climbers, hikers, and guides. Often private services work in cooperation with voluntary services. For instance, a private helicopter rescue team may work with a volunteer mountain-rescue team on the ground. Mountain rescue is often still free, although more and more states and localities are considering charging for it. In Switzerland mountain rescue is highly expensive and will be charged to those rescued. In many remote or less-developed parts of the world, organized mountain rescue services do not exist.

Oregon has a charge-back law that allows localities to recover costs of a rescue. Rescue teams typically oppose such laws. The premise of such laws is deterring risky behavior by making those who get lost pay for their rescue or making their families pay for recovery of their bodies. At least five states, including Oregon and California, have "charge-for-rescue" laws on the books. But the Mountain Rescue Association, which represents about 100 volunteer groups in the United States, Canada, and Britain, strongly objects to the concept. Their concern is based on the behavior of lost hikers. If people believe they are going to be charged, especially a big charge, they are going to be afraid to summon help. They will try and get themselves out of the problem. They will delay, which can make the difference between life and death. Wilderness advocates, however, argue that the purpose and values of the national park system encourage self-reliance.

Climbing accidents are decreasing. In 2005 the average annual number of reported climbing accidents had declined from a peak of 168 in the 1980s to 159 in the 1990s and 139 so far this decade. Injuries fell from 146 to 128 to 117 in those decades. Average annual deaths peaked at 34 in the 1970s, then dropped to 29 in the 1980s, 27 in the 1990s, and 26 so far this decade. Mountain rescues do not cost much, in comparison to other outdoor emergencies. Most climbing rescues are performed by highly skilled volunteer rescue units who do not charge, or by specialized park rangers whose costs are often subsidized by climbing use fees. This makes climbing rescues less of a cost on taxpayers than rescues of other recreational users. When military helicopters are used, the rescue operations can double as training exercises and be covered under existing programs.

POTENTIAL FOR FUTURE CONTROVERSY

As population increases and recreational trips to mountains and other dangerous natural places increase, more rescues are foreseeable. Technology in terms of locators, personal early warning systems, and communication systems help decrease mortality and allow corpse recovery. This technology is also expensive, with helicopters costing about $10,000 a day. As costs increase because of those who may be taking unnecessary risks in nature, there is a strong push to make them pay for their own rescues.

See also Ecotourism as a Basis for Protection of Biodiversity; Federal Environmental Land Use

Web Resources

American Alpine Club. National Policy Issues. Available at position www.americanalpine club.org/pages/page/32. Accessed January 21, 2008.

Oregon Mountaineering Association. Don't Put a Price on Rescue. Available at www.i-world. net/oma/news/rescue/oregonian-2002–06.html. Accessed January 21, 2008.

Types of Rescues. Available at www.answers.com/topic/search-and-rescue. Accessed January 21, 2008.

Further Reading: Salkeld, Audrey. 2005. *World Mountaineering*. New York: Sterling Publishing.

MULTICULTURAL ENVIRONMENTAL EDUCATION

Environmental education is a process that promotes the analysis and understanding of environmental issues and questions as the basis for effective education, problem solving, policy making, and management. It has also been the focus of a growing number of state and federal environmental policies. The purpose of environmental education is to educate individuals so they possess the knowledge and skill to effectively communicate knowledge to both children and adults. In the broader context, environmental education's purpose is to assist in the development of a citizenry conscious of the scope and complexity of current and emerging environmental problems and supportive of solutions and policies that are ecologically sound. To impart the capacity to engage in public participation is also a civic goal, and part of the emerging civic environmentalism. The controversies of the environment, as unresolved as they are today, need to be taught in a multicultural context with many different perceptions of "environment."

The perception of environment, and risk from external experiences with the environment, varies greatly by gender, race, income, and education. White males have the lowest perception of risk; all females and all nonwhite people share about the same perception of risk. Urban children of color count the built environment and everything around them as "environment," while suburban white children tend to view open fields lined with forests as "environment." The failure to appreciate these differences in perception of risk from environment underlies many of the controversies in U.S. environmental policy. This is especially noted in the area of public involvement and participation, and increased citizen involvement in environmental and land-use decisions. The vast majority of Environmental Protection Agency (EPA) and state environmental agency personnel and industry environmental managers are white males. White males have been shown to have an aura of invincibility about environmental risk perception. In some research in this controversial area of risk perception by race and gender, the greater the education and income of the white male the more likely they were to agree with putting communities at risk without disclosure. Many of the older risk models were based on a 150-pound white male, one of the healthiest and least vulnerable segments of U.S. society. Risk perception and risk models were skewed to one demographic, one that is not characteristic or shared by an increasing majority of people. Multicultural educators often see no choice for environmental education to be anything but multicultural because of their teaching environments. Experiential learning activities in the environment of urban areas can expose areas of toxic waste accumulation, environmental injustice, and pollution. This is also a battleground within this controversy. Is it risky for children to engage in these activities? What if they find out some controversial environmental information? Environmental education in the culture of a rural logging community with mill closures, of Native American lands and reservations, of farmworkers, and around military bases faces different contextual battlegrounds, and the clash of cultures around environmental issues.

THE NATIONAL ENVIRONMENTAL EDUCATION ACT (NEEA) OF 1990

The NEEA was a controversial piece of legislation that created a national policy of environmental education. It helped environmental education gain a foothold in the curriculum of universities and education generally. It also represented the merger of environmental and educational policies. Some object because they contend that the students are too young to decide complicated issues and are misled into one-sided and anti-industrial views. Many educators like to use environmental education to spark interest in its scientific foundations. Students, like citizens, now have the right to public information that was never before available, such as the Toxics Release Inventory. The following findings led to the historic passage of this act and also laid the foundation for state approaches.

FINDINGS

1. Threats to human health and environmental quality are increasingly complex, involving a wide range of conventional and toxic contaminants in the air and water and on the land.
2. There is growing evidence of international environmental problems, such as global warming, ocean pollution, and declines in species diversity, and these problems pose serious threats to human health and the environment on a global scale.
3. Environmental problems represent as significant a threat to the quality of life and the economic vitality of urban areas as they do to the natural balance of rural areas.
4. Effective response to complex environmental problems requires understanding of the natural and built environment, awareness of environmental problems and their origins (including those in urban areas), and the skills to solve these problems.
5. Development of effective solutions to environmental problems and effective implementation of environmental programs requires a well-educated and trained professional workforce.
6. Current federal efforts to inform and educate the public concerning the natural and built environment and environmental problems are not adequate.
7. Existing federal support for development and training of professionals in environmental fields is not sufficient.
8. The federal government, acting through the Environmental Protection Agency, should work with local educational institutions, state educational agencies, not-for-profit educational and environmental organizations, noncommercial educational-broadcasting entities, and private-sector interests to support development of curricula, special projects, and other activities, to increase understanding of the natural and built environment, and to improve awareness of environmental problems.
9. The federal government, acting through the coordinated efforts of its agencies and with the leadership of the Environmental Protection Agency, should work with local educational institutions, state educational agencies, not-for-profit educational and environmental organizations, noncommercial educational-broadcasting entities, and private-sector interests to develop programs to provide increased emphasis and financial resources for the purpose of attracting students into environmental engineering and assisting them in pursuing the programs to complete the advanced technical education required to provide effective problem-solving capabilities for complex environmental issues.
10. Federal natural resource agencies such as the U.S. Forest Service have a wide range of environmental expertise and a long history of cooperation with educational institutions and technology transfer that can assist in furthering the purposes of the act.

POLICY

It is the policy of the United States to establish and support a program of education on the environment, for students and personnel working with students, through activities in schools, institutions of higher education, and related educational activities, and to encourage postsecondary students to pursue careers related to the environment.

Because schools are the first to show changing demographics in education they are more multicultural. Relating to learners on their own terms in ways they understand is fundamental to the pedagogic mission of elementary and secondary education. For environmental education, itself sometimes controversial, to be effective in its mission, it must adapt to a multicultural society.

ENVIRONMENTAL EDUCATION

The following are guiding principles for environmental education.

- Consider the environment in its totality, both natural and built: biological and physical phenomena and their interrelations with social, economic, political, technological, cultural, historical, moral, and aesthetic aspects.
- Integrate knowledge from the disciplines across the natural sciences, social sciences, and humanities.
- Examine the scope and complexity of environmental problems and develop critical-thinking and problem-solving skills and the ability to synthesize data from many fields.
- Develop awareness and understanding of global problems, issues, and interdependence, helping people to think globally and act locally.
- Consider both short-and long-term futures in matters of local, national, regional, and international importance.
- Emphasize the role of values, morality, and ethics in shaping attitudes and actions affecting the environment.
- Stress the need for active citizen participation in solving environmental problems and preventing new ones.
- Enable learners to play a role in planning their learning experiences and provide an opportunity for making decisions and accepting their consequences.
- Develop the need for lifelong learning beginning at preschool level and continuing throughout formal elementary, secondary, and postsecondary levels and utilize nonformal modes for all age and education levels.

Environmental education does not advocate a particular viewpoint or course of action. Rather, environmental education teaches individuals how to weigh various sides of an issue through critical thinking, and it enhances their own problem-solving and decision-making skills.

POTENTIAL FOR FUTURE CONTROVERSY

As U.S. society becomes more multicultural, and as populations generally increase, our environmental impacts also accumulate. Solutions to some environmental problems can come from shared and collective perceptions of the environmental problem at hand. These can come from scientists, community activists, government officials, and industry leaders.

Our children will see and suffer from environmental impacts we have ignored. The consequences of environmental actions from the past now can occur within a lifetime. Drastic, controversial, and rough actions may be necessary for environmental sustainability. The scientists, community residents, government officials, and industry leaders of the future are the children of today.

Multicultural environmental education will continue as a controversy because it teaches about environmental controversies, and because of the clashes and differences in perceptions of the environment by class, race, and gender. Because of the experiential character of environmental education, some educators may find themselves embroiled in environmental controversy. Some schools themselves may be dangerous to children if the water is polluted, or if the school was built on top of a landfill. When communities discover an environmental risk through their children, they quickly react. This increases the citizen involvement in environmental decision making and public participation if available. It may also increase the ultimate transparency in environmental reporting and decision making. The new multicultural generation of environmentalists asks questions that challenge the accountability of both government and industry. That challenge will also be a battleground.

See also Community-Based Science; Public Involvement and Participation in Environmental Decisions; Sustainability; Toxics Release Inventory

Web Resources

Association for Environmental and Outdoor Education. Diversity in Outdoor/Environmental Education. Available at aeoe.org/resources/diversity/index.html. Accessed January 21, 2008.

Multicultural Education Internet Resource Guide. Available at jan.ucc.nau.edu/~jar/Multi.html#3. Accessed January 21, 2008.

U.S. Environmental Protection Agency. Environmental Education: Guidelines and Assessment Tools. Available at www.epa.gov/enviroed/guidelines.html. Accessed January 21, 2008.

Further Reading: National Environmental Education Advisory Council. 2005. *Setting the Standard, Measuring the Results, Celebrating Successes: A Report to Congress on Environmental Education in the US*. Washington, DC: U.S. Environmental Protection Agency.

N

NANOTECHNOLOGY

Both environmental controversy and hope are attached to emerging nano-technology. Promises of fast and thorough cleanups, highly efficient water purification systems, and very low emissions and pollution from industrial manufacturing processes accompany concern that there may be too much unregulated risk.

WHAT IS NANOTECHNOLOGY?

Nanotechnology is difficult to define in terms of policy, practices, fields, or disciplines. Fundamentally, it is the manipulation of atoms at the molecular level. The molecules can come from different elements. They are very, very small, between 0.1 and 100 nanometers. One nanometer is one billionth of a meter. One result of this small size is that some of the fundamental properties of materials can change. Clusters of gold and silver show catalytic properties when otherwise inert, for example. Many smaller molecules have much greater surface area and reactivity.

HOW DO YOU MAKE MATERIALS AT THE NANOTECHNOLOGICAL LEVEL?

This is where many current research efforts are focused. Science has done basic research, and engineering applications are being explored. There are three ways to produce nanomaterials currently. They can be put together one mole-cule at a time. Some nanoparticles are easier to work with than others. Another method, one that fuels research interest in self-replication, is the ability of some molecules to self-assemble. Carbon nanotubes of 60 atom carbon molecules can form much like snowflakes do. The third method is to remove nanomaterials from larger particles. All these methods are experimental. The search is on for useful applications of this technology, and some of them are explicitly environmental.

APPLICATIONS OF NANOTECHNOLOGY

Nanotechnology is touted as the next paradigm shift in human technological advancement. It is currently used in some computer components, car parts and catalytic converters, scratch-resistant coatings and paints, antibacteriological bandages and socks, sunscreens and other cosmetics, and self-cleaning windows. Possible future nanotechnological applications are to dramatically increase the efficiency of solar energy cells, quickly and efficiently clean up toxic wastes and pollution, find and destroy inoperable cancers, and develop combat suits that morph camouflage and absorb bullet impacts, more efficient fuel cells, highly efficient water-purification processes, and self-replicating nanobots.

Nanotechnology will revolutionize many aspects of current life, its advocates say. It does offer a very broad platform for applications in industry, biomedicine, and the environment. It has also altered the political economy of scientific research by reversing a trend toward hyperspecialization. International interest in nanotechnology is very high. More than 60 countries have invested in research and development on nanotechnology, expending roughly eight billion dollars in 2005. The rate of research and development is so fast that it is outpacing the ability of government to regulate it. There has been little assessment of short-or long-term public health or environmental impacts. Proponents of the precautionary principle would argue that this is an excellent case against its application because the risk is powerful and could be irreversible. Nanotechnology advocates claim that devices using nanotechnology will be lighter, smaller, and faster in their applications and will use less raw materials and energy. The decrease in environmental impacts at the manufacturing stage and the decreased waste indicate a decreased overall environmental impact. Others vehemently disagree and want the industry regulated to prevent negative environmental impacts.

BATTLEGROUNDS: NOT ENOUGH REGULATION FOR THE RISK?

Environmentalists and others fear that commercialization of nanotechnology is outpacing knowledge about possible risk to public health and the environment. There is national and international concern about this, and battlegrounds are forming. The concern is that these tiny molecules could enter the human body via ingestion, inhalation, or dermatological absorption. If they are powerful they could also be dangerous once inside the human body. Small particulate matter and soot in air pollution is known to cause damage to the developing lungs of children. What if the nanotubes are carcinogenic? There are very few animal studies on that point. A study in 2003 suggested that nanotubes may damage the lungs of exposed rats and mice. Another study in 2004 suggested that nanotubes can accumulate in some organs, in this study causing brain damage in exposed fish.

Nanotechnology industry proponents point out that these are animal, not human studies. They suggest that these particles are too small to be damaging. When germ theory was first introduced to medicine, some doctors rejected it outright. How could something as small as a germ cause so much damage and death, they reasoned at that time. The industry has proposed to regulate itself, creating one of the first large battlegrounds. The federal government has also created a nanotechnology initiative.

In 2007, a group called Environmental Defense, a self-described conservation group, and Dupont Corporation developed a framework for industry self-regulation. The Dupont Corporation manufactures chemicals and many other products, and has done so for many years. Together they developed the Nano Risk Framework. Environmentalists and others point out that this is a private industry initiative proposing self-regulation. Nuclear energy was self-regulated until major accidents threatened the public. Industry is concerned that they will lose a competitive advantage due to overcautious environmental regulation.

The U.S. nanotechnology initiative has promoted the development of the technology and identified unresolved occupational safety research needs. One major concern is how workers in the industry would be exposed. With many toxic materials, the workers are the most exposed. There is also concern about impacts and risks of nanotechnology. Another safety research issue is how would nanotubes contaminate an environment? They can be made from different materials. Are some materials better in terms of safety? Are some so dangerous they should be banned? These enormous gaps in knowledge challenge all stakeholders.

There is also a battleground brewing internationally around nanotechnology. The United Nations Educational, Scientific, and Cultural Organization (UNESCO) is concerned about the ethics of scientific research around nanotechnology. They believe that the rate of development is so rapid that an anticipatory approach to ethical questions is necessary.

POTENTIAL FOR FUTURE CONTROVERSY

There is a large controversy brewing about regulation of nanotechnology. U.S. industries do not want to lose any competitive edge, nor does any nation. The U.S. Environmental Protection Agency is just beginning to examine some of the possible risk scenarios. Releasing a powerful, self-replicating technology into the environment creates uneasiness among environmentalists and some citizen and labor groups. They recall when the United States allowed the nuclear energy industry to be self regulated and major accidents, such as Three Mile Island, threatened people.

See also Precautionary Principle; Solar Energy Supply; Sustainability

Web Resources

Center for Biological and Environmental Nanotechnology (CBEN) at Rice University. Available at http://cnst.rice.edu/cben/. Accessed January 21, 2008.

National Nanotechnology Initiative (NNI). Available at www.nano.gov/. Accessed January 21, 2008.

U.S. Environmental Protection Agency. National Center for Environmental Research: Nanotechnology: Sensors. Available at es.epa.gov/ncer/nano/research/nano_sensors.html. Accessed January 21, 2008.

Further Reading: Goldman, Lynn, and Christine Coussens. 2005. *Implications of Nanotechnology for Environmental Health Research.* Washington, DC: National Academies Press; National Science and Technology Council (U.S.) Subcommittee on Nanoscale Science, Engineering, and Technology. 2001. *Societal Implications of Nanoscience and Nanotechnology.* New York: Springer; Roco, Mihail C., and William Sims Bainbridge. 2006. *The Nanotech Pioneers: Where Are They Taking Us?* Weinheim, Germany: Wiley-VCH; Schwarz, James A., Cristian I. Contescu, and Karol Putyera. 2004. *Dekker Encyclopedia of Nanoscience and Nanotechnology.* Boca Raton, FL: CRC Press; Sung Hee, Joo I., and Francis Cheng. 2006. *Nanotechnology for Environmental Remediation.* New York: Springer.

NATIONAL PARKS AND CONCESSIONS

The U.S. national park system has been controversial since its inception and is often involved in cutting-edge environmental controversies. An historic and modern controversy is the extent to which concessions are granted to businesses to operate in the park. Concessions include many uses that have environmental impacts. Some environmentalists do not want any concessions to business operations in national parks.

National parks are predominantly in the western United States, part of the vast tracts of public lands there. Significant national parks and monuments exist in every state. An old park service policy of granting concession monopolies, without open bidding, turned the minds of many against national parks in general. Some of the parks also operated their own concessions in the form of lodging, guides, and so forth at that time. In many areas the national parks provided needed jobs and tourist revenue, especially during the Great Depression of the 1930s. The national parks still do provide this function as a revenue corridor to the local economy as long as they offer what tourists want in an accessible way. Tourist activities like using all-terrain vehicles and off-road driving in general, snowmobiling, river rafting, and skiing all impact the environment and may therefore conflict with the overall mission of the national park system.

In the many jurisdictional controversies around the fear of federal encroachment on states' rights, the issue of concessions to local residents and their businesses was one of the compromises necessary for the acceptance of a strong federal presence. Arkansas, Oklahoma, Wyoming, Montana, Washington, Colorado, California, and Oregon have ceded jurisdiction over their national parks to the federal government. With a mission to protect the species and ecosystems within their boundaries, as part of strong federal land-use planning in general, national parks are still a battleground for concessions.

One example of some of these dynamics is Rocky Mountain National Park in Colorado. One of the oldest national parks and long a battleground for concessions, this park has beautiful mountains. This makes it attractive for the lucrative ski market that Colorado is noted for. Rocky Mountain's problems with winter sports development stemmed from several sources. Park Service philosophy maintained that all outdoor sports, including winter sports, should be encouraged. Also, some powerful park administrators believed that to get appropriations from a parsimonious Congress they had to publicize the recreational potential of the park system. Others contended that visitors should be allowed to use their parks to the fullest no matter what the environmental impacts. As a result, ski lifts were eventually built in Mt. Rainier, Sequoia, Yosemite, Lassen Volcanic and Olympic national parks. To implement these directives in Rocky Mountain, without marring the scenery, became the special problem of more than one superintendent, but the fact that a winter sports complex was built gave evidence of Park Service appeasement of local political pressures.

CAMPGROUNDS

As in several other parks, Rocky Mountain officials in Rocky Mountain have been bothered by the presence of inholdings and campgrounds. The existence of both was considered ecologically unsound, since the environment for wildlife was irrevocably altered. Thus it was a sound practice to buy out privately developed lands in the park. To replace them with campgrounds was a politically realistic policy around this type of concession. Pressures from politicians and chambers of commerce demanding more campgrounds, roads, and trails are an ever-present concern to the park administrators at Rocky Mountain. Car travel had already greatly changed the character of concessions required by the public. No longer could hotels within the park compete profitably with campgrounds within and motels outside of the park's boundaries.

ENVIRONMENTAL IMPACTS OF CONCESSIONS: NOISE

Human-made noise in national parks and public lands is a battleground. Controversy and litigation have increased in parks where recreational park users hear touring planes and helicopters, snowmobiles, watercraft, off-road vehicles, and even the National Park Service's own equipment and concessions. Members of environmental groups, off-highway-vehicle groups, the air-tourism industry, tribal nations, and some of the major government agencies that oversee public lands all have major issues with noise control, but not always the same ones. Large recreational vehicles often need to run generators, so large campgrounds have more noise impact. Some communities want to expand their airports to take economic advantage of the presence of a national park. Larger airports mean bigger planes and more tourist revenue. It also means extending the environmental impacts of noise and air pollution in the surrounding vicinity. Some wonder what the land-use limits with concessions will be in the future. Will the national parks allow racecar driving, manufacturing industries, or tall office buildings? Noise issues with concessions are a developing battleground.

a result of this accident. The range of environmental impacts of the Chernobyl meltdown is a battleground. It is involved with ecosystem risk assessment, faulty long-term monitoring, and the cleanup priorities and practices. The catastrophic risk of a nuclear meltdown with an unknown and controversial range of human and ecological impacts underscores modern tension around nuclear energy.

THE UNITED STATES TODAY

Today, about 103 nuclear reactors are operating in 31 states. They generate about one-fifth of the nation's electricity. Major expansions are planned, and each will be a battleground for this controversy. With this growth comes a much closer scrutiny of the environmental costs and benefits of nuclear energy by environmentalists, government agencies, and competing energy sources.

One battleground for alternative energy sources is the marketplace. Today's market forces support nuclear power. The electricity industry is being deregulated, allowing consumers and their cities to avoid the forced contracts of hydroelectric power companies. Existing nuclear plants appear to be a low-cost, alternative energy source. Many power plants run on coal or petrochemicals, with high levels of emissions into the air. This has a powerful impact on global warming and climate change. A main cause of climate change, global warming, air pollution, and acid rain is carbon dioxide emissions. Nuclear reactors do not emit any carbon dioxide. Industry proponents tout the new and improved safety of modern plants to alleviate regulator and consumer fears. In the United States, electricity from nuclear power plants was greater than that from oil, natural gas, and hydropower. Coal accounts for 55 percent of U.S. electricity generation. Nuclear plants generate more than half the electricity in at least six states. According to industry statistics, average operating costs of U.S. nuclear plants dropped substantially during the 1990s. Expensive downtime, for maintenance and inspections, has been steadily reduced. Licensed commercial reactors generated electricity at a record-high average of more than 87 percent of their total capacity in 2000, which indicates increasing demand. In the battleground of the marketplace, nuclear energy is gaining international and domestic appeal. However, environmentalists and site communities remain concerned. They are concerned about human and environmental impacts due to exposure from employment, transit of nuclear waste, spills, and other environmental sources. Nuclear environmental impacts are some of the most powerful ones humans can create, and they can destroy any resiliency in a given ecosystem. They last a very long time and can move through the soil and water to contaminate other parts of an ecosystem. Radiation may remain unstable and lethal for 100,000 years. Nuclear waste is currently stored in holding pools and casks alongside the power plants. Some have expressed concern with leaking casks. Radiation is a potential problem in every phase, from mining the uranium, to operating the plant, and finally disposing of the waste. Low-level radioactive waste is also a pressing environmental controversy. Cleaning up severe environmental problems at U.S. nuclear weapons production facilities alone is expected to cost at least $150 billion over the next several decades. Cleaning up old nuclear energy plants is another large expense. Each is followed by community controversy about exposure and adequacy of cleanup.

NEW POWER PLANTS: NEW BATTLEGROUNDS

Because of the powerful environmental impacts of nuclear energy this controversy will persist. There is as yet no solution to the waste problem. Old power plants generate public concern about safety. Building new plants will be expensive. If recent history is a reliable indicator, cost overruns can be expected that will impact the price of electricity. Also expect community resistance that can effectively block new nuclear power generators. Community resistance can take the form of refusing to finance any aspect of design, construction, or operation. When the Washington Public Supply System tried to build five nuclear power plants during the mid-1970s, environmental lawsuits for violations of the required environmental impact statements, and community resistance to take or pay contracts from the Bonneville Power Administration, led to the plan's collapse. More than $3 billion of default on taxpayer bonds then occurred, resulting in 43 lawsuits in five states. Many investors lost substantial sums of money. The courts were clogged with long, complicated cases about municipal finance, as well as the environmental lawsuits that followed the project.

THERMAL POLLUTION CONTROVERSY

Thermal pollution, the addition of heated water to the environment, is a type of pollution that recently has come to public attention. In England the largest single industrial use of water is for cooling purposes, while in the United States in 1964, 49,000 billion gallons of water were used by industrial manufacturing plants and investor-owned thermal electric utilities. Ninety percent, or 44,000 billion gallons, of this amount was used for cooling or condensing purposes primarily by electric utilities. With the increased demand for greater volumes and less expensive electric power, the power companies are rapidly expanding the number of generating plants, especially nuclear-powered plants.

To them, nuclear plants offer many advantages over conventional plants but have one major drawback seriously affecting the environment, which is excess-heat losses. These plants are only 40 percent as efficient as conventional plants in converting fuel to electricity; that and loss of efficiency manifests itself as waste heat. As the number of nuclear power plants and other industrial plants increase, an estimated ninefold increase in waste-heat output will result.

Nuclear plant liquid releases fall into the following categories:

- Nonradioactive
- Slightly radioactive

Water that has been used to cool the condenser and various heat exchangers used in the turbine-generator support processes, or that has passed through the cooling towers is considered nonradioactive. The cooling towers remove heat from the water discharged from the condenser so that the water can be discharged to the river or recirculated and reused.

What water goes through the cooling towers differs from plant to plant. Some nuclear power plants use cooling towers as a method of cooling the circulating water that has been heated in the condenser. Nuclear powers plants also differ in when they emit hot water into the environment. During colder months, the discharge from the condenser may be directed to the river. Recirculation of the water back to the inlet to the condenser occurs during certain fish-sensitive times of the year, during which the nuclear power plant is supposed to limit its thermal emissions. Many environmentalists contend that they do not do so and that even when they do the environmental impacts of hot water on the aquatic ecosystem are too severe. The thermal emissions of a nuclear plant are powerful and can heat up large bodies of water. They can heat the circulating water as much as 40°F. Some nuclear power plants have placed limits on the thermal differential allowed in their coolant water emissions. For example, they may have limits of no more than 5°F difference between intake and outflow water temperatures. Cooling towers essentially moderate the temperature to decrease the thermal impact in the water but also decrease power plant efficiency because the cooling-tower pumps consume a lot of power.

Some or all of this water may be discharged to a river, sea, or lake. One way to reduce thermal pollution is to make use of the hot water and steam using cogeneration principles.

Usually water released from the steam generator is also nonradioactive. Less than 400 gallons per day is considered low leakage and may be allowed from the reactor cooling system to the secondary cooling system of the steam generator. This can be a battleground because it creates concern that radioactivity will seep out. By law, where radioactive water may be released to the environment, it must be stored and radioactivity levels reduced below certain levels. These levels themselves can be another battleground. Citizens frequently challenge experts over nuclear risk issues.

In terms of the environmental impacts of thermal pollution, much remains unstudied. Water that is too warm can damage endangered species, such as some types of salmon. This thermal pollution causes a variety of ecological effects in the aquatic ecosystem. More must be learned about these effects to ensure adequate regulation of thermal discharges.

Industry proponents claim that the small amounts of radioactivity released by nuclear plants during normal operation do not pose significant hazards to workers, the community, or the environment. What concerns many communities is the potential for long-term hazardous waste disposal. There could be deadly cumulative effects. There is scientific uncertainty about the level of risk posed by low levels of radiation exposure. Problems inherent in most risk assessments, such as failing to account for population vulnerability or dose-response variance, do little to assure communities they are safe. Human health effects can be clearly measured only at high exposure levels, such as nuclear war. Other human health effects are generalized from animal studies. In the case of radiation, the assumed risk of low-level exposure has been extrapolated from health effects among persons exposed to high levels of radiation. Industry proponents argue that it is impossible to have zero exposure to radiation because of low levels of background radiation. There is public and community concern about the cumulative impacts of radiation generally.

INDUSTRY AND GOVERNMENT RESPONSIBILITY

Because of the high level of public concern, there are strict protocols for safety. Responsibility for nuclear safety compliance lies with nuclear utilities that run the power plants and self-report most of the environmental information. By law, they are required to identify any problems with their plants and report them to the Nuclear Regulatory Commission (NRC). These reports, and the lack of them, have been battlegrounds. Nuclear power plants last about 40 years and then must be closed in a process called *decommissioning*. The NRC requires all nuclear utilities to make payments to special trust funds to ensure that money is available to remove all radioactive material from reactors when they are decommissioned. Several plants have been retired before their licenses expired, whereas others could seek license renewals to operate longer. Some observers predict that more than half of today's 103 licensed reactors could be decommissioned by the year 2016. There may be a battleground looming as to whether there is enough money in the trust funds to clean up the sites adequately. The decommissioning of these power plants will be a battleground because of the controversies surrounding the disposal of low-level radioactive waste. It will also be very expensive and fraught with scientific uncertainty. By law, the federal government is responsible for permanent disposal of commercial spent fuel and federally generated radioactive waste. Choosing sites for this waste is a battleground. States have the authority to develop disposal facilities for commercial low-level waste. This is a battleground for many states. The siting process for these types of waste sites is itself a battleground on the local level, often engaging strong community protests. Generally the federal government can preempt state authority, which can preempt local authorities. In this battleground, lawsuits are common.

NUCLEAR WASTE: IS THERE A SOLUTION?

One of the battlegrounds of nuclear power is the disposal of the radioactive waste. It must be sealed and put in a place that cannot be breached for thousands of years. It may not be possible to make warning signs that last long enough. Thousand-year timescales are well beyond the capability of current environmental models. A whole range of natural disasters could ensue and breach the waste site. Few sites can withstand an earthquake or volcanic eruption. Wastes are stored on-site, then moved to a waste transfer station, then to a terminal hazardous waste site. There are battlegrounds at each step in the process. Each nuclear reactor produces an annual average of about 20 tons of highly radioactive spent nuclear fuel and 50–200 cubic meters of low-level radioactive waste. Over the usual 40 year permits granted to nuclear power plants by the NRC, this amounts to a total of about 800 tons of radioactive spent fuel and 1,000–8,000 cubic meters of low-level radioactive waste. There are additional hazardous materials used in the operation of the power plant. Upon decommissioning, contaminated reactor components are also disposed of as low-level waste. When combined with any hazardous waste that was stored on the site, the waste produced can be quite large.

The cradle-to-grave exposure to radiation, the increased regulation of hazardous vehicle routes in cities, and the likely expansion of nuclear energy to more community sites all portend a larger controversy.

POTENTIAL FOR FUTURE CONTROVERSY

As climate change becomes a more salient political issue, the push for nuclear energy becomes stronger. There is still no policy to deal with the dangerous waste this energy process produces, which is a source of growing controversy. Scientific controversies about dose-response levels with radiation exposure, cancer causation, and effects on vulnerable populations close to communities all continue. Environmentalists have traditionally opposed nuclear energy as a source of power, but some groups have recently begun to question this in light of global warming and greenhouse gas emissions from coal and oil sources.

See also Acid Rain; Cancer from Electromagnetic Radiation; Cumulative Emissions, Impacts, and Risks; Ecosystem Risk Assessment; Low-Level Radioactive Waste; Sustainability; Toxic Waste and Race

Web Resources

Environmental Literacy Council. Energy. Available at www.enviroliteracy.org/category. php/4.html. Accessed January 21, 2008.

Reaching Critical Will. Nuclear Energy. Available at www.reachingcriticalwill.org/resources/ factsheets/energy.html. Accessed January 21, 2008.

Further Reading: Bodansky, David. 2004. *Nuclear Energy: Principles, Practices, and Prospects.* New York: Springer; Collin, Robert W. 1989. "Creditors' Remedies in Municipal Default: The Washington Public Power Supply Cases." In *State and Local Government Debt Financing*, ed. M. David Gelfand. New York: Callahan and Co. Publishing; Garwin, Richard L., and Georges Charpak. 2002. *Megawatts and Megatons: The Future of Nuclear Power and Nuclear Weapons.* Chicago: University of Chicago Press; Hodgson, Peter Edward. 1999. *Nuclear Power, Energy and the Environment.* London: Imperial College Press; Kuletz, Valerie L. 1998. *The Tainted Desert: Environmental Ruin in the American West.* New York: Routledge; Zwaan, B.C.C. van der, C. R. Hill, and A. L. Mecheynk, eds. 1999. *Nuclear Energy: Promise or Peril?* Singapore: World Scientific.

O

ORGANIC FARMING

As concern over the safety of pesticides increases, more people turn to organic foods. Food is considered organic when it is grown in soil without chemical or pesticide contamination, not treated with pesticides while growing, and not treated with preservatives after harvest. Each stage of the food process is a battleground. The array of labels changes, confusing consumers and contributing to this controversy.

WHAT IS SO GREAT ABOUT ORGANIC?

As new rules and laws develop new battlegrounds over what is organic, consumers seek organic produce because they distrust chemicals. The idea is that by growing a natural product as close to how nature intended it as possible consumers will decrease their risk of chemical exposure. The increased consumer demand for organic products is in part a market representation of the level of distrust of chemicals, especially preservatives and pesticides. The raging battleground is how much unsafe chemical exposure results from vectors like residues and crop treatments.

Thorough pesticide-residue testing done by the U.S. Department of Agriculture (USDA) found that conventional fresh fruits and vegetables are:

- three to more than four times more likely on average to contain residues than organic produce;
- 8 to 11 times more likely to contain multiple pesticide residues than organic samples; and

- shown to contain residues at levels 3 to 10 times higher, on average, than corresponding residues in organic samples.

Providing organic fruits and vegetables gives a choice proven to significantly reduce dietary exposure to pesticides. Some nonorganic foods in the United States are heavily contaminated with pesticides. Some of these foods are frequently consumed by infants and children:

Fruits	Vegetables
Apples	Celery
Cherries	Spinach
Peaches	Sweet Bell Peppers
Pears	
Nectarines	
Strawberries	

According to the USDA, multiple pesticide residues are commonly found in these nine fruits and vegetables. The pesticide risk reduction benefits of consuming organic apples, pears, peaches, strawberries, cherries, celery, spinach, and sweet bell peppers were found to be particularly significant, especially for woman of childbearing age and infants and children. Some of the chemicals used in nonorganic foods can cross the placental barrier and negatively affect the fetus. It is a scientific battleground as to which chemicals create specific harm, and which ones can migrate to the fetus. Industry claims that many of these chemicals pose no actual risk to humans, especially if the fruit and vegetables are washed prior to consumption. They point out there are only animal studies and that the doses were far larger than humans are exposed to via food. Public health advocates claim mothers and unborn children are very vulnerable and that mothers would prefer to err on the side of caution with their child. Public health advocates also point out that industry claims ignore the environmental reality of human exposure by failing to account for cumulative risks.

The importation of organic goods raises another battleground. Agricultural industries are very sensitive to trade tariffs and import/export controls. They want to control their markets and protect their crops. Many nations have no environmental controls on the fruits and vegetables they produce. They can use pesticides that are illegal in the United States, and in quantities that persist a long time. The pesticide residues found by the USDA in imported organic samples show that they are six times riskier than the pesticide residues found in domestic organic samples. This battleground eventually produced federal legislation designed to control these risks. The National Organic Program (NOP) requires all imported produce to meet the requirements of the NOP in order to be sold as organic in the United States. One major and persistent battleground is the certification process. Standards change and are often contested.

ORGANIC CERTIFICATION

Organic food consumers are very discerning about food content. The label organic can open new markets for farmers and grocers, as well as increase the prices. Food certified as organic can have different meanings. Organic certification establishes and confirms standards. Consumers want to know everything they can about what they consume. Organic certification gives consumers confidence in the food they purchase and consume.

Shifting standards only add to organic farming controversies but are necessary as science, technology, and consumer demand bear down on food. These standards are specific to land, produce, and animals.

Land that has been used for industry or agriculture may have chemical loads in the soil and water. This is called *conventional* land. The land itself must be organic for the products to be organic. To be certified as organic the land used to grow the produce must be free of chemicals for at least three years before the

---

LABELS AND THEIR MEANING

Tags on organic fruits and vegetables start with a nine and have five numbers whereas conventional ones have four numbers. Genetically modified crops also have five numbers but start with the number eight. The main terms are defined as follows.

Conventional—product or produce is made conventionally with commonly used pesticides or chemicals.

Transitional—producer is working toward certification and is in the three-year period where they are meeting organic standards in practice but cannot be certified until the land has been in transition to organic status for at least three years.

100 percent organic—exclusively organic ingredients or single-ingredient products.

Organic—95 percent of the product is made of organic ingredients (excluding salt and water).

Made with organic ingredients—ingredients are 70–95 percent organic; organic ingredients are specified on the label.

Less than 70 percent organic—product has some organic ingredients but less than 70 percent.

Natural—There are no artificial flavors, colors, or preservatives in that product and it is minimally processed.

Free-range—This means that the animals have some space to roam. This does not always mean outdoor pasture.

Hormone-free—The animals were not given hormones, e.g., growth hormones; usually applies to dairy cows and cattle.

Currently, many hormones, additives, and warehouse exposure vectors are not included in the label. Labeling is also a regulatory act that needs enforcement when neglected, violated, or misleading.

crops are grown. If there is still a chemical presence after a three-year period, then another two years may generally be required. During the three-year period, the land is considered transitional. Animals are likewise strictly controlled. Organic animal husbandry often uses free-range animal production. The animals are allowed to walk freely as they grow. One question is how free do they have to be to be free range? Many nonorganic commercial meat-production operations have no minimum required space per animal. Pigs, chickens, goats, cattle, and others all require a minimum amount of land, depending on food sources. Organic poultry and beef are raised on organic feed or organic pastures. Organic cows and chickens cannot be given growth hormones, stimulants, or antibiotics. Organic dairy pasteurization must meet hygienic standards. There is a subbattleground with the sale of whole, unpasteurized milk. Some consumers want it, but it is illegal to sell it for public health reasons. Some organic farmers sell it anyway. Dairy cows may not be sold for meat. Beef is processed in a certified plant where organic cows are separated from conventional cattle.

In all these categories, organic standards are still evolving. This can cause controversy. To add to the confusion, labeling requirements differ. The general trend is to include all the ingredients on the label. Organic food consumers have a higher expectation of this. Battlegrounds around labeling of organic foods include organic as well as the content listings. Should organic food have to list ingredients that conventional foods do not? Should conventional foods list trans fat? Should milk labeling disclose whether the cow that produced it was given growth hormones?

CONTROVERSIES IN ORGANIC FARMING OPERATIONS IN THE UNITED STATES

Although much of the battle between conventional food and organic food is fought in the marketplace, it is also fought in the arena of regulations and their enforcement. It is a competitive advantage to have food labeled as organic, but if farmers are increasing productivity through pesticides and also labeling it organic, they are competing unfairly. Enforcement of the rules is very difficult in agricultural operations and promises to be a contentious issue.

Pesticides sometimes drift in the air from a nonorganic field to an organic field. Pesticide drift can be an intense and costly battleground between neighboring agricultural operations. Organic crop fields are vulnerable to pesticides in the environment from other sources. Once a field is retainted with chemicals, it could take years to bring it back into production with the organic certification. In agribusiness industrial operations, drift losses less than 10 percent are rare. The pesticide operations are large, delivered by airplanes (crop dusters) over large areas. Pesticide drift has been so bad at times that it has closed interstate highways. Farmworkers and the environment are also affected. Irrigation water is another potential source of pesticide drift onto organic farms. There is some scientific controversy on this point. Scientists maintain that because of the dilution in water it is unlikely that pesticide contamination of organic crops by irrigation water is a significant problem. Organic farmers contend that dilution is not as effective as theoretically thought when drought, other sources of water pollution, and increasing numbers of other water users impede dilution. Environmentalists have long argued that dilution of chemicals in the environment is not a solution if there is environmental damage along the way to dilution.

NATURAL PESTICIDES?

A second cluster of controversies surround the use of natural pesticides on organic farms. There is concern over the risks they may pose the consumer. The National Organic Program has approved a number of pesticides containing natural ingredients for use by certified organic farmers. Residues of some of these natural substances are common whether produced on an organic or conventional farm. This small battleground in this controversy will continue to fester.

POTENTIAL FOR FUTURE CONTROVERSY

With the rapid expansion of food-production efficiency after World War II came the rise of large agricultural corporations. Combined with grants of land given to universities for agricultural research (known as land grant colleges), the green revolution produced large amounts of food where there was none before. The scale was larger, the use of chemicals extensive, and the distribution networks longer. The environmental impact is enormous, the mixture of chemicals in the soil now unknown, and the energy costs of shipping goods long distances expensive. Organic farmers are much more than gardeners with an environmental conscience. As concern about long-term sustainability emerges as a societal goal, organic farming emerges as a way to practice what one preaches.

This controversy has potential to grow. Demand for organic produce increases as the cost of conventional produce increases. Those who can afford to, buy organic over nonorganic when given the choice. As more land comes into production for organic food, more efficient distribution networks will develop. Organic food is already a big business. Part of its value lies in full and complete labeling of food. Labeling requirements and rule enforcement will grow as consumers, public health advocates, and organic food industries seek more accurate and accessible food information.

See also Cumulative Emissions, Impacts, and Risks; Farmworkers and Environmental Justice; Hormone Disruptors: Endocrine Disruptors; Pesticides; Sustainability

Web Resources

California Certified Organic Farmers. Available at www.ccof.org/. Accessed January 21, 2008.

World Wide Opportunities on Organic Farms. Available at www.wwoof.org/. Accessed January 21, 2008.

Further Reading: Degregori, Thomas R. 2004. *Origins of the Organic Agriculture Debate.* Ames, IA: Blackwell Publishing; Lipson, Elaine Marie. 2001. *The Organic Foods Sourcebook.* New York: McGraw-Hill; Organisation for Economic Co-operation and Development. 2003. *Organic Agriculture: Sustainability, Markets, and Policies.* Washington, DC: Organisation for Economic Co-operation and Development.

P

PERMITTING INDUSTRIAL EMISSIONS: AIR

The U.S. Environmental Protection Agency (EPA) requires large emitters of some chemicals to get permits to do so. Permits are required only if the emissions exceed a certain threshold. Emissions are self-reported by industry. Once a permit is issued, it places limits on the chemicals that can be emitted into the air. There are many exceptions, weak governmental enforcement, and many lawsuits. Early lawsuits by environmental groups were successful in forcing the EPA to develop clean air standards.

REGULATORY BACKGROUND

The Clean Air Act gave authority to the EPA to issue clean air standards, which it did for some emissions after environmental groups sued them under special provisions called citizen suits. Controversies involve weak enforcement of environmental regulations that allow too many emissions. Much of this controversy occurs in the courts and at the EPA. The rules and regulations around industrial air emissions are complex but do allow for citizen input. Courts require many plaintiffs to exhaust administrative remedies before going to court. Therefore, it is necessary to know how permits basically work in the regulation of industrial emissions into the air.

Congress established the New Source Review (NSR) permitting program as part of the 1977 Clean Air Act amendments. The NSR is a preconstruction permitting program that serves two important purposes. First, it ensures that air quality is not significantly degraded from the addition of new and modified factories, industrial boilers, and power plants. In areas with unhealthy air, NSR assures that new emissions do not slow progress toward cleaner air. In areas with clean air, especially pristine areas like national parks, NSR assures that new emissions do not significantly worsen air quality. Second, the NSR program assures people that any large new or modified industrial source in their neighborhoods will be as clean as possible, and that advances in pollution control occur concurrently with industrial expansion. The NSR permits are legal documents that the facility owners/operators must abide by. The permit specifies what construction is allowed, what emission limits must be met, and often how the emissions source must be operated.

There are three types of NSR permitting requirements. A source may have to meet one or more of these permitting requirements. They are:

- Prevention of significant deterioration (PSD) permits that are required for new major sources or a major source making a major modification in an attainment area;
- Nonattainment NSR permits that are required for new major sources or major sources making a major modification in a nonattainment area; and
- Minor source permits.

WHAT ARE PERMITS?

Permits are legal documents that the industry must follow. They specify what construction is allowed, what emission limits must be met, and often how the source must be operated. They may contain conditions to make sure that the source is built to match parameters in the application that the permit agency relied on in their analysis. For example, the permit may specify stack heights that the agency used in their analysis of the source. Some limits in the permit may be there at the request of the source to keep them out of other requirements. For example, the source may accept limits in a minor NSR permit to keep the source out of PSD. To assure that sources follow the permit requirements, permits also contain monitoring, record-keeping, and reporting requirements.

WHO ISSUES THE PERMITS?

The EPA issues the permit in some cases. State and local air pollution control agencies may have their own permit programs that are approved by the EPA in the State Implementation Plan (SIP), or they may be delegated the authority to issue permits on behalf of the EPA. Most NSR permits are issued by state or local air pollution control agencies. The EPA establishes the basic requirements for an NSR program in its federal regulations. States may develop unique NSR requirements and procedures tailored for the air quality needs of each area as long as the program is at least as stringent as EPA's requirements. A state's NSR program is defined and codified in its SIP. In some cases, state or local air pollution control agencies have not developed a unique NSR program and rely completely on EPA's NSR program. These states are delegated the authority to issue permits on behalf of the EPA and are often referred to as *delegated states*. Finally, the EPA is the permitting authority in some areas. In both delegated programs and where the EPA issues permits, the rules and procedures followed in issuing NSR permits are specified in EPA regulations.

PREVENTION OF SIGNIFICANT DETERIORATION (PSD)

Prevention of significant deterioration applies to new major sources or major modifications at existing sources for pollutants when the area where the source is located is in attainment or unclassifiable with the National Ambient Air Quality Standards (NAAQS).

The Clean Air Act, which was last amended in 1990, requires the EPA to set NAAQS for widespread pollutants from numerous and diverse sources considered harmful to public health and the environment. The Clean Air Act established two types of national air quality standards. Primary standards set limits to protect public health, including the health of sensitive populations such as asthmatics, children, and the elderly. Secondary standards set limits to protect public welfare, including protection against visibility impairment and damage to animals, crops, vegetation, and buildings. The Clean Air Act requires periodic review of the science upon which the standards are based and the standards themselves. This process can be controversial within the scientific community.

The EPA has set NAAQS for six principal pollutants, which are called *criteria* pollutants.

Prevention of significant deterioration permits require the following:

1. Installation of the Best Available Control Technology (BACT). BACT is an emissions limitation that is based on the maximum degree of control that can be achieved. It is a case-by-case decision that considers energy, environmental, and economic impact. BACT can be add-on control equipment or modification of the production processes or methods. This includes fuel cleaning or treatment and innovative fuel combustion techniques. BACT may be a design, equipment, work practice, or operational standard if imposition of an emissions standard is infeasible.

2. An air quality analysis. The main purpose of the air quality analysis is to demonstrate that new emissions from a proposed major stationary source or major modification, in conjunction with other applicable emissions increases and decreases from existing sources, will not cause or contribute to a violation of any applicable NAAQS or PSD increment. Generally, the analysis will involve an assessment of existing air quality, which may include ambient monitoring data and air quality dispersion modeling results, and predictions, using dispersion modeling, of ambient concentrations that will result from the applicant's proposed project and future growth associated with the project.

3. An additional impacts analysis. The additional impacts analysis assesses the impacts of air, ground, and water pollution on soils, vegetation, and visibility caused by any increase in emissions of any regulated pollutant

Table P.1 Pollution Emissions and Trading Them

Pollutant	Primary Stds.	Averaging Times	Secondary Stds.
Carbon Monoxide	9 ppm (10 mg/m³)	8-hour[1]	None
	35 ppm (40 mg/m³)	1-hour[1]	None
Lead	1.5 µg/m³	Quarterly Average	Same as Primary
Nitrogen Dioxide	0.053 ppm (100 µg/m³)	Annual (Arithmetic Mean)	Same as Primary
Particulate Matter (PM$_{10}$)	Revoked[2]	Annual[2] (Arith. Mean)	
	150 µg/m³	24-hour[3]	
Particulate Matter (PM$_{2.5}$)	15.0 µg/m³	Annual[4] (Arith. Mean)	Same as Primary
	35 µg/m³	24-hour[5]	
Ozone	0.08 ppm	8-hour[6]	Same as Primary
	0.12 ppm	1-hour[7] (Applies only in limited areas)	Same as Primary
Sulfur Oxides	0.03 ppm	Annual (Arith. Mean)	- - - - - - -
	0.14 ppm	24-hour[1]	- - - - - - -
	- - - - - - -	3-hour[1]	0.5 ppm (1300 µg/m³)

[1] Not to be exceeded more than once per year.

[2] Due to a lack of evidence linking health problems to long-term exposure to coarse particle pollution, the agency revoked the annual PM$_{10}$ standard in 2006 (effective December 17, 2006).

[3] Not to be exceeded more than once per year on average over three years.

[4] To attain this standard, the three-year average of the weighted annual mean PM$_{2.5}$ concentrations from single or multiple community-oriented monitors must not exceed 15.0 µg/m³.

[5] To attain this standard, the three-year average of the 98th percentile of 24-hour concentrations at each population-oriented monitor within an area must not exceed 35 µg/m³ (effective December 17, 2006).

[6] To attain this standard, the three-year average of the fourth-highest daily maximum eight-hour average ozone concentrations measured at each monitor within an area over each year must not exceed 0.08 ppm.

[7] (a) The standard is attained when the expected number of days per calendar year with maximum hourly average concentrations above 0.12 ppm is ≤ 1, as determined by appendix H. (b) As of June 15, 2005, the EPA revoked the one-hour ozone standard in all areas except the 14 eight-hour ozone nonattainment Early Action Compact (EAC) Areas.

from the source or modification under review and from associated growth. Associated growth is industrial, commercial, and residential growth that will occur in the area due to the source.

4. Public involvement. Prior to issuing a permit, permitting authorities generally follow these steps:

1. The state agency determines whether the permit application is complete enough to begin processing it.
2. They publish a notice to inform the public of the public comment period (usually 30 days), and the deadline for requesting a public hearing on the draft permit.

The notice can be published in a newspaper of general circulation in the area where the source is located or in a state publication, like a state register. Some agencies also post the public notice and other information on their Web site. They decide whether to revise the draft permit based on the comments received. In some cases the permitting authority may publish a notice and seek comments on the revised permit. They then usually issue the permit.

WHAT IS PSD'S PURPOSE?

Prevention of significant deterioration does not prevent sources from increasing emissions. Instead, PSD is designed to:

1. protect public health and welfare;
2. preserve, protect, and enhance the air quality in national parks, national wilderness areas, national monuments, national seashores, and other areas of special national or regional natural, recreational, scenic, or historic value;
3. ensure that economic growth will occur in a manner consistent with the preservation of existing clean air resources; and
4. assure that any decision to permit increased air pollution in any area to which this section applies is made only after careful evaluation of all the consequences of such a decision and after adequate procedural opportunities for informed public participation in the decision-making process. There can be controversy at this juncture.

COMMUNITY ACTIONS

Many communities turn to their government for help in fighting air pollution, for example, enforcement actions against sources that are not complying with their permits. Is there anything a citizen can do?

You can notify the permitting authority or the EPA if you believe a facility did not obtain an NSR permit before construction or is not complying with its permits. The EPA may refer you to the appropriate state or local agency that handles the type of violation you are reporting. An environmental violation occurs when an activity or an existing condition does not comply with an environmental law or regulation. Environmental violations can include (but are not limited to):

- Smoke or other emissions from local industrial facilities;
- Tampering with emission control or air conditioning systems in automobiles;
- Improper treatment, storage, or disposal of hazardous wastes;
- Exceedances of pollutant limits at publicly owned wastewater treatment plants;
- Unpermitted dredging or filling of waters and wetlands;
- Any unpermitted industrial activity;
- Late-night dumping or any criminal activity including falsifying reports or other documents.

An environmental emergency is a sudden threat to the public health or the well-being of the environment, arising from the release or potential release of oil, radioactive materials, or hazardous chemicals into the air, land, or water. Examples of environmental emergencies include:

- Oil and chemical spills
- Radiological and biological discharges
- Accidents causing releases of pollutants

Section 304 of the Clean Air Act allows citizens to sue to enforce many of the Clean Air Act's requirements. Lawsuits may be filed against the source, the state permitting authority, and the EPA.

Suggested Strategy for Reviewing a Title V Permit

- Step One: Identify and Locate the Underlying Source for Any Requirement Mentioned in the Permit Application or Draft Permit
- Step Two: Review the Permit Application for Helpful Information
- Step Three: Review the Statement of Basis
- Step Four: Review General Conditions
- Step Five: Check to See if Source-Specific Air Quality Requirements Are Correctly Applied to the Facility
- Step Six: Check to See Whether Any Federal Requirements Are Incorrectly Identified as State-Only Requirements

Many communities also get involved at permit-renewal and permit-modification phases. Some additional conditions in permits required by some states include records of any compliance actions taken against them. Even though many environmental advocates complain that actual notice of permit changes is poor, the issue is increasingly important to communities. Public computers make public access to notices better.

HOW CAN I COMMENT ON NSR PERMITS?

As a member of the public, you can use the NSR program to ensure that sources are complying with the requirements that apply to them. NSR gives you the opportunity to:

1. Comment on and request a public hearing on permits before they are issued.
2. Appeal permits issued pursuant to the State Implementation Plan (SIP). The appeal procedures will depend on the state the source is located in. For state-specific information, get in touch with the appropriate contact listed on your state's permit contact page, which is available by clicking on your state on the Where You Live page U.S. map (www.epa.gov/epahome/whereyoulive.htm).

3. Appeal or permits issued by the EPA or by state or local agencies that are issuing the permit on behalf of the EPA to the Environmental Appeals Board and the federal courts.

Normally one must have commented on the draft permit to appeal it.

Communities and environmentalists complain that this notice is often ineffective in reaching people and the process is ineffective in listening to people. Industry is often concerned that public participation in these forums exposes them to potential citizen lawsuits. The Clean Air Act specifically allows citizen lawsuits. Communities and environmentalists often want to monitor plant emissions. Industry strongly discourages this because of fear of exposure to liability. Citizen monitoring often begins in these controversial cases.

POTENTIAL FOR FUTURE CONTROVERSY

Industrial air emissions are an important controversy that evokes strong emotions from communities. They are concerned about the health of their families. Now, due in part to the Toxics Release Inventory, communities are more empowered with knowledge. Environmental organizations have a long history of successful litigation against the EPA in enforcing the Clean Air Act and show no signs of letting up. Industry is incompletely regulated based on self-reported data. Emissions from all sources are unknown. Concerns about cumulative effects and societal interest in sustainability mean that industrial air emissions will continue to be controversial.

See also Air Pollution; Childhood Asthma and the Environment; Citizen Monitoring of Environmental Decisions; Cumulative Emissions, Impacts, and Risks; Ecosystem Risk Assessment; Incineration and Resource Recovery; Pollution Rights or Emissions Trading; Toxics Release Inventory

Web Resources

U.S. Environmental Protection Agency. Compliance and Enforcement. Available at www.epa.gov/compliance/. Accessed January 21, 2008.

U.S. Environmental Protection Agency. New Source Review (NSR): Where You Live. Available at www.epa.gov/nsr/where.html. Accessed January 21, 2008.

U.S. Environmental Protection Agency. Technology Transfer Network: Clean Air Technology Center. Available at cfpub1.epa.gov/rblc/htm/bl02.cfm. Accessed January 21, 2008.

U.S. Environmental Protection Agency. Where You Live: Environmental Violations. Available at www.epa.gov/epahome/violations.htm. Accessed January 21, 2008.

Further Reading: Bradstreet, Jeffrey W. 1996. *Hazardous Air Pollutants: Assessment, Liabilities, and Regulatory Compliance.* New York: William Andrew Inc; Gonzalez, George A. 2005. *The Politics of Air Pollution: Urban Growth, Ecological Modernization, and Symbolic Inclusion.* New York: SUNY Press; Greenway, A. Roger. 2002. *How to Obtain Air Quality Permits.* New York: McGraw-Hill Professional; Trzupek, Richard. 2002. *Air Quality Compliance and Permitting Manual.* New York: McGraw-Hill Professional.

PERMITTING INDUSTRIAL EMISSIONS: WATER

Some industries and cities discharge wastes and chemicals into the water. This discharge is mixed with nonpoint sources of pollution such as runoff from paved areas. Many cities have old sewage overflow systems that were supposed to discharge directly into the water when it stormed. Populations grew, but infrastructural maintenance did not keep up. Population and pavement increases have increased the flow and toxicity of urban wastewater streams. Communities blame industry, past and present, for water pollution, but industry blames to nonpoint sources. Meanwhile, the EPA is very slow at issuing some of the standards like the total maximum daily load of chemicals a given watershed can carry. Environmentalists can sue under the Clean Water Act and do so, claiming weak enforcement of the law by the EPA. All the while, some of the discharges are accumulating as sediment, citizens are increasingly monitoring the water quality themselves, and water quantity and quality are decreasing.

The water controversies spill over to every area of environmental policy, especially if ecosystem or ecological risk assessment approaches are used. All approaches to sustainability must include water. The focus of this environmental controversy is industrial discharges into water.

The pollution of water has a serious impact on all living organisms and can negatively affect the use of water for drinking, household needs, recreation, fishing, transportation, and commerce for a long time. Pollution can accumulate on river bottoms from the beginning of industrialization. The EPA enforces federal clean water and safe drinking water laws, provides support for municipal wastewater treatment plants, and takes part in pollution-prevention efforts aimed at protecting watersheds and sources of drinking water.

CLEAN WATER ACT HISTORY

The Clean Water Act dominates water pollution control. It is a fairly recent policy development and is strewn with battlegrounds at almost every point of implementation. Growing public awareness and concern about water pollution led to enactment of the Federal Water Pollution Control Act Amendments of 1972. As amended in 1977, this law became known as the Clean Water Act. It gave the EPA the authority to implement pollution-control programs such as setting wastewater standards for industry. The Clean Water Act also continued to set standards for all regulated contaminants in surface waters. There is an ongoing dispute as to whether all contaminants are included in the regulatory reach. The act made it unlawful for any person to discharge any pollutant from a point source into navigable waters, unless a permit was obtained under its provisions. Once a permit is obtained they may discharge the pollution into the water as long as they do not exceed the limit in the permit. However, discharges are often averaged out with a rolling average that can be used to keep water discharges within the limit. Environmentalists and some downstream communities strongly challenge this practice because it allows degradation of the aquatic environment to occur. The Clean Water Act also funded the construction of many sewage treatment plants. Without this help, many communities would be without wastewater treatment.

Nonpoint source (NPS) pollution, unlike pollution from industrial and sewage treatment plants, comes from many diffuse sources. NPS pollution is caused by rainfall or snowmelt moving over and through the ground. As the runoff moves, it picks up and carries away natural and human-made pollutants, finally depositing them into lakes, rivers, wetlands, coastal waters, and even our underground sources of drinking water. These pollutants include:

- Excess fertilizers, herbicides, and insecticides from agricultural lands and residential areas;
- Oil, grease, and toxic chemicals from urban runoff and energy production;
- Sediment from improperly managed construction sites, crop and forest lands, and eroding streambanks;
- Salt from irrigation practices and acid drainage from abandoned mines; and
- Bacteria and nutrients from livestock, pet wastes, and faulty septic systems.

Atmospheric deposition and hydromodification are also sources of nonpoint source pollution.

States report that nonpoint source pollution is the leading remaining cause of water quality problems. The effects of nonpoint source pollutants on specific waters vary and may not always be fully assessed. These pollutants have harmful effects on drinking water supplies, recreation, fisheries, and wildlife. Nonpoint source pollution results from a wide variety of human activities on the land. Some activities are federal responsibilities, such as ensuring that federal lands are properly managed to reduce soil erosion. Some are state responsibilities, for example, developing legislation to govern mining and logging and to protect groundwater. Others are handled locally, such as by zoning or erosion-control ordinances.

FEDERAL RESPONSIBILITIES

Federal laws require government approval prior to beginning any work in or over navigable waters of the United States that affects the course, location, condition, or capacity of such waters, or prior to discharging dredged or fill material into the waters of the United States. Regulatory programs that implement these laws are administered through permits issued by the U.S. Army Corps of Engineers, but it shares responsibility with the Environmental Protection Agency (EPA).

WETLANDS

Wetlands are very important parts of ecosystems that often have suffered large environmental impacts. Although many states have their own wetland regulations, the federal government bears a major responsibility for regulating wetlands. Each federal agency has a different mission that is reflected in the agency's legal authority for wetland protection. The Army Corps' duties to wetlands are related to navigation and water supply; they control activities like dredging, channelizing large rivers, and constructing levees. All these activities can become site-specific battlegrounds, often ending in federal court. The EPA's legal responsibilities are related to protecting wetlands primarily for their contributions to the quality of water for drinking and swimming. Only 20 percent of surface water is even monitored at all, and it has experienced a steady drop in water quality. More watersheds are being monitored. Community awareness of water monitoring and testing methods has increased. The Toxics Release Inventory has further strengthened citizens and their participation in water-permitting processes. The Fish and Wildlife Department's legal duties are related to managing fish and wildlife-game species and threatened and endangered species. They have a very strong interest in the ecological integrity of wetlands. States are becoming more active in wetland protection. Many states have adopted programs to protect wetlands beyond those programs enacted by the federal government.

Conflicting interests of state and federal agencies, and with private property owners, create many site-specific battlegrounds with permitting industrial uses of water.

In a 2000 report to Congress, the EPA cited nonpoint sources of pollution as the top factors making the remaining 40 percent of the nation's waterways too polluted for swimming or fishing. The Clean Water Act only regulates point sources. As scientists and environmental activists recognized the value of wetlands in mitigating pollution, the EPA began to emphasize wetlands protection under the Clean Water Act.

PROVISIONS CREATE CONTROVERSY

Under the Clean Water Act, the EPA sets national water quality criteria and specifies levels of various chemical pollutants that are allowable under these criteria. The discharge of regulated chemicals into surface waters is controlled by the National Pollutant Discharge Elimination System (NPDES), which requires polluters to obtain federal permits for every chemical they discharge. The permits, which can be issued by the EPA or by state government agencies, give a business or municipality the right to discharge a limited amount of a specific pollutant. The NPDES has been criticized by industry groups for ambiguous regulatory policies and long delays in granting permits. Environmentalists and some community groups counter that the delays are because the pollution is a great concern. They do not want it.

The EPA also took steps toward cleaning up polluted waterways and regulating nonpoint source pollution in 2000. The agency introduced new rules that encouraged individual states to identify dirty waterways and establish standards to help eliminate sources of pollution. The states were required to come up with a maximum amount of pollution that each waterway could absorb. This measurement was known as the total maximum daily load (TMDL). Then the states had to decide which local landowners or businesses needed to

reduce their pollution levels to meet the TMDL. The states were also required to evaluate future development plans near the waterways to make sure they would not increase pollution levels. It soon became clear that the TMDL program would be very controversial, as it has been every time it has come close to implementation. Some cities seeking industrially based economic development and industry trade groups maintain that that these provisions discourage development along already-polluted waterways and restrict the rights of property owners. Others were concerned that compliance with the new regulations would be too expensive in terms of time and cash outlays for new equipment. Often, if there is no room left in a water permit, the waste must be shipped to a hazardous waste landfill. It may have to be treated before it can go there. This is a very expensive process.

WATER PERMITS FOR LAND DEVELOPMENT

Section 404 Dredge and Fill Permits of the Clean Water Act created a special permitting program to regulate discharge of dredged and fill material into wetlands and other waters of the United States. The Army Corps of Engineers is principally responsible for issuing permits under this program. Many of the controversies over 404 permits involve whether the discharge area qualifies as a wetland. In addition, the act exempts some discharges of dredge and fill material from the regulations. Exempt activities can each be a battleground and include:

- normal farming, forestry, and ranching activities;
- maintenance and reconstruction of many water structures (dams, dikes, etc.);
- construction and maintenance of farm and forest roads; and
- activities associated with certain state-approved programs.

WETLANDS

After a Supreme Court decision in 2001 cut back federal jurisdiction over wetlands, federal agencies under the George W. Bush administration in January 2003 proposed new rules to follow up on and carry out that decision. Wetlands provide many benefits: flood control, water purification, groundwater recharge, migratory resting areas for birds and butterflies, feeding and breeding habitat for fish and wildlife, erosion control, and recreation. They mitigate the impacts of many natural disasters. The Pacific Coast Federation of Fishermen's Associations estimated that almost $79 billion per year was generated from wetland-dependent species in 1997, or about 71 percent of the nation's entire $111 billion commercial and recreational fishing industry.

Wetlands are environmentally degraded by many human activities. Commercial and residential development is one of the human activities that destroy wetlands. Wetlands produce rich soil and are often cleared and drained for farmland. Road construction, river channelization, and dams can destroy wetlands. Earlier attitudes toward wetlands revolved around draining them for the public health (especially because of mosquitoes), treating them as

garbage and waste sites, and letting all uses develop there no matter what their environmental impact. Other wetlands are lost to natural resource extraction like mining, logging, and grazing. The nature of wetlands varies according to area, as do the environmental impacts and risks to them. The sea-level rise expected to accompany global warming will also destroy some coastal wetlands.

POTENTIAL FOR FUTURE CONTROVERSY

As concern over water quality and quantity increases so too will controversy over how the government protects the water from environmental impacts. Concern about cumulative effects will force examination of all known sources. Industries with permits stand out although other sources of pollution exist, such as sewage from cities and runoff from land development. Waste is increasing as populations increase, and land development and paving continue unabated. At the same time social concern over cumulative effects, sustainability, and global warming is increasing. Industries that are currently required to be permitted and many cities will try to resist permit requirements. Much of the information is self-reported by industry to government. Environmentalists complain that enforcement of law is weak and advocate in court. Industry complains that the federal agencies vacillate in their interpretation of the law. Communities want strict enforcement of the law, transparent environmental transactions, and complete accountability for any and all discharges.

The battleground for this controversy will move back and forth between legislatures and the courts as policy makers grapple with the thorny issues of clean water.

MERCURY EXPOSURE FROM THE FISH IN THE GREAT LAKES

The primary means of human exposure to mercury is consumption of fish containing methyl mercury, so groups who eat Great Lakes fish as a major source of nutrition are the most at-risk populations. These groups include low-income African Americans, Native Americans, and non-English speakers such as some Hmong immigrants. The benefits and burdens to these stakeholders are complicated. Reduction of mercury levels is a benefit because it provides access to clean air, water, and food. Stakeholders also face significant burdens because strict regulation of fishing could cut off a major source of sustenance.

One concern is that present and future generations lack social and environmental education. Specifically, many are unaware of the undetectable dangers of toxic chemicals and how to recognize the health effects of contamination. Even if the knowledge would not benefit fishers, women, and children, who perceive and experience different levels of risk, need information about mercury concentrations in fish and the risks of contamination.

Individual stakeholders may assess their risks in different ways, depending on physical and socioeconomic variables as well as acceptance or awareness of the problem. Benefits

and burdens may not be entirely clear without individual stakeholders speaking for themselves, and total informed consent is integral to fair decision making. Total informed consent is impossible if stakeholders are not properly educated about the present and future effect of social and environmental issues.

For example, the community needs to understand that the state of Michigan's standard regulating point-source discharge of mercury assumes a fish consumption rate of 6.5 grams per person per day, which is inadequate, because it fails to represent consumption levels of some minority groups. This lack of information increases the likelihood of mercury poisoning in groups who consume more fish due to lack of warning.

Lack of information about contaminated fish and lack of access to uncontaminated fish jeopardizes the community's health and culture as well as the health of future generations.

Groups such as environmentalists and evangelicals also have an interest in the mercury issue. For example, the leader of the National Association of Evangelicals is eager to reframe mercury regulations as a pro-life issue because "curbing mercury emissions protects children from learning disabilities and unborn children from brain damage—that gets people's attention."

Environmental groups are interested stakeholders, speaking on behalf of silent stakeholders such as flora and fauna in the Great Lakes basin. There is no formal program for collecting data on the impacts of mercury on the ecology of the Great Lakes basin; however, risks have been clearly identified for birds, fish, and mammals. Mercury has had a documented effect on birds, such as loons and birds of prey in eastern North America; at least one loon death in Nova Scotia has been linked directly to the toxin. Mercury-related deaths and illnesses indicate that the flora and fauna in the basin would benefit from decreased levels of mercury in their environment.

Coal-fired power plants account for about 40 percent of total U.S. human-made mercury emissions. The Clean Air Mercury Rule is specifically targeted at power plants because of their large share of overall emissions. The utility industry has a stake in the EPA's mercury rule because it determines the timeframe and extent of industry's compliance. The benefits and burdens to the coal-based power industry are reflected by trade associations, which exist solely to pursue the greatest good for their members. According to the Edison Electric Institute (EEI), the benefits of the Clean Air Mercury Rule are ease of implementation, economic security, energy security, environmental gains, flexibility, and technology incentives. The organization lists no burdens. The potential burdens to industry of stricter controls include the cost of implementing more efficient technologies or the threat of economic losses due to a switch to alternative energy sources. However, because of the EPA's calculations, these burdens have been mitigated. One benefit of stricter controls, based on the EEI's list, is that they are more likely to receive environmental gains.

By Monica Patel, JD Lewis and Clark Law School, 2006

See also Climate Change; Cumulative Emissions, Impacts, and Risks; Global Warming; Sustainability; Total Maximum Daily Load (TMDL) of Chemicals in Water; Water Pollution

Web Resources

National Council for Science and the Environment. Controversies over Redefining "Fill Material" under the Clean Water Act. Available at www.ncseonline.org/NLE/CRS/abstract. cfm?NLEid=218. Accessed January 21, 2008.

U.S. Environmental Protection Agency. Envirofacts Data Warehouse: Water. Available at www.epa.gov/enviro/html/pcs/pcs_overview.html#PCS. Accessed January 21, 2008.

Further Reading: Greenway, A. Roger. 2004. *How to Obtain Water Quality Permits.* New York: McGraw-Hill Professional; Ryan, Mark. 2004. *The Clean Water Act Handbook.* Chicago: American Bar Association; Wiersma, Bruce G. 2004. *Environmental Monitoring.* Washington, DC: CRC Press.

PERSISTENT ORGANIC POLLUTANTS

Persistent organic pollutants present environmental and health risks despite their effectiveness in their uses and applications. There is a strong international movement to ban them from food sources, but some countries still use them, and chemical manufacturing corporations still produce them for profit.

WHY ARE PERSISTENT ORGANIC POLLUTANTS A CONTROVERSY?

These chemicals cause controversy because they last a long time in the environment. Their presence can be damaging to other parts of the soil and water. Since they persist, or last, in the environment, annual reapplication of pesticides increases cumulative exposures dramatically. Persistent organic pollutants (POPs) are several groups of chemicals. Polychlorinated biphenyls (PCBs) are industrial chemicals. The two remaining groups are dioxins and furans. POPs have one common characteristic, their persistence in the environment. Some early pesticide applications wanted this characteristic because it was presumed that stronger chemicals that lasted longer performed their task better. They last longer than required for their intended use, however, and it is always a battleground as to exactly how long they do last. Over time the accumulation of POPs eventually made the case that they do persist. All 12 POPs listed are chlorinated compounds, nine of them having been developed as pesticides. Their use is decreasing, but controlling international use is controversial. Farmworkers and others who live near POP applications can suffer from overexposure. This can cause acute poisoning. If acute poisoning occurs, no antidotes are available for the internationally banned POPs. POP exposure can follow other vectors of exposure because they are so persistent in the environment. They eventually reach the top of the food chain—humans.

HUMAN EXPOSURE

The greatest part of human exposure to the listed POPs comes from the food chain. The contamination of food, including breast milk, by POPs is a worldwide controversy. In most of the world, breast milk is the sole source of food for most infants. Edible oils and animal products are most often involved in case of POP contamination. Food contaminated by POPs can pose chronic health risks, including cancer. The long-term implications of low-level exposure are not known. There is controversy on this point on the scientific battlegrounds about POPs. Some researchers are concerned that long-term low-level exposure to POPs may have more cumulative impacts because of their persistence. Others maintain that low-level exposures do not cause any risk but do not engage the cumulative-risk concerns.

Vectors for food contamination by POPs occur through environmental pollution of the air, water, and soil, or through the use of organochlorine pesticides. Food contamination by POPs can have a significant impact on food exports and imports. At the international level, limits for residues of persistent organochlorine insecticides have been established for a range of commodities. They are recognized by the World Trade Organization as the international reference in assessing nontariff barriers to trade. Because of this, international bodies are major players in the controversies over POPs.

DISPOSAL OF POPs

Most countries are facing the problem of disposal of some remaining POPs. This is a large controversy because of the cost of doing so and the environmental and public health risks of not doing so.

The strict requirements for proper disposal of these chemicals create an enormous burden for a developing country and their industries, both economically and technologically. Legal aspects of transboundary movement of POPs are very specialized and time-consuming. The temptation to illegally dispose of POPs can be strong.

DO POPS EVER DEGRADE NATURALLY?

There have been recent claims that POPs can degrade naturally. Some say it is a type of bioremediation. If this is the case, the cost of cleanups decreases dramatically because POPs can be left to degrade in place. Environmentalists generally prefer bioremediation because it usually has lower environmental impacts.

The controversy over whether POPs can be naturally degraded by microbial action is a long-standing one. New research indicates that this occurs for DDT. Research also indicates that naturally occurring organisms in sediments play an important role in breaking down the chlorinated compounds. The finding that DDE, a toxic by-product of the pesticide DDT, can naturally degrade comes from laboratory experiments performed by researchers from Michigan State University's Center for Microbial Ecology. They used marine sediments collected from a Superfund site off the coast of southern California. Their research samples came from the Palos Verdes Shelf, the subject of one of the largest

Natural Resources Damage Assessment cases in the United States. More than 20 years after they were deposited, DDT compounds are still present in surface sediments at levels harmful to life. But according to the Michigan State University microbiologist, the experiments

> do not prove dechlorination is taking place at a significant rate in the sediments at the site. They do demonstrate that there are sediment microbes that can dechlorinate what was previously considered a terminal product.

The EPA's most likely plan for the Superfund site is to cover part of the ocean floor with a cap of thick sand, a project that could cost as much as $300 million.

POTENTIAL FOR FUTURE CONTROVERSY

The POPs list is likely to expand as our use and knowledge of them increases. So too will the list of potential alternatives to some POPs. Eliminating them from food chains and human breast milk will be a big first step when it eventually happens, but other more inclusive policy approaches will be more controversial. Waste disposal, bioremediation, cost of cleanup and who pays for it all remain debated and growing areas of environmental policy. The battleground for the POPs controversy will remain in the international environmental community via treaties and international bodies like the United Nations. The battleground for mandatory disposal is just beginning and could shape the cleanup policies of the host country. Environmentalists fear an increase in illegal ocean dumping.

PERSISTENT ORGANIC POLLUTANTS

Polychlorinated Biphenyls (PCBs) in Hudson River, New York

General Electric (GE) is responsible for PCBs on the river bottom of the Hudson River in New York. It is an old company that existed well before any environmental regulation. GE contends that natural processes, including reductive dechlorination, have substantially reduced the risk to humans and the environment and that these processes should be allowed to continue. The U.S. Environmental Protection Agency is considering a Superfund cleanup of the contaminated sediments, which would make GE a primary responsible party. They can then either clean it up according to EPA specifications, or the EPA will do it and charge them for the expense. Environmentalists have several lawsuits engaged in this controversy. Community groups along the historic Hudson River are very concerned.

The EPA concluded that dechlorination will not naturally remediate contaminated sediments. According to GE a number of natural processes, when viewed together, dramatically reduce the risk from contaminated sediments: "Anaerobic dechlorination reduces toxicity; aerobic degradation reduces the overall mass; and sorption onto organic particles reduces bioavailability."

Hudson River PCBs are a serious health risk, according to community and environmental groups.

- PCBs can damage the immune, reproductive, nervous, and endocrine systems. They can impair children's physical and intellectual development.
- PCBs cause cancer in animals and are strongly linked to human cancer, according to studies by leading health agencies.
- GE says PCBs do not hurt people, citing a study it commissioned on workers at its Hudson Falls plant. The New York State Department of Health and many independent scientists critiqued the research and said it does not support GE's claims.
- According to the EPA, cancer risks from eating upper-river PCB-contaminated fish are 1,000 times over its goal for protection and 10 times the highest level generally allowed by Superfund law.

Hudson River PCBs will not go away naturally. There is deep distrust from the community that they are safe from harm from these PCBs.

- PCBs were designed not to break down. They are persistent organic pollutants that remain in the environment indefinitely.
- GE claims river microbes eliminate PCBs naturally, but the EPA found that less than 10 percent have broken down. After breakdown, PCBs remain toxic and are more readily spread throughout the ecosystem.
- GE claims Hudson River PCB pollution has dropped 90 percent, a deceptive statistic because the drop occurred when discharges were banned in the late 1970s. Since the mid-1980s, levels have remained quite constant and well above acceptable limits. The EPA's independent, peer-reviewed science predicts the problem will last into the foreseeable future without remediation.
- GE's PCBs are responsible for eat-none health advisories for women of childbearing age and children for all fish from all Hudson River locations.

Hudson River PCBs are not safely buried by sediments, contend community members and scientists. This is a pervasive controversial issue in environmental cleanups. Does disturbing the site cause more environmental damage? Often this question is complicated by cost-cutting measures that affect the environmental decision. The early days of "don't ask, don't tell" environmental policy are replaced by ecosystem risk assessment at Superfund sites. In the case of the Hudson River and GE,

- Of the estimated 1.3 million pounds of PCBs dumped by GE, about 200,000 pounds remain in upper-river sediments. Every day, through resuspension by currents, boats, bottom-dwelling animals, and so on, the sediments release PCBs. About 500 pounds wash over the Troy Dam annually.
- The EPA's peer-reviewed science has found that PCBs are not being widely buried by sediments.

Current Dredging Technology Is Safe, Effective, and Efficient

- Dredging will cut in half the flow of PCBs over the Troy Dam, and the EPA forecasts safe fish levels 20 years earlier after dredging.

- The EPA's proposal does not rely on a local landfill.
- Under the EPA's worst-case scenario, dredging might stir up 20 pounds of PCBs annually. However, the cleanup will immediately and dramatically reduce the 500 pounds moving downstream already. In the long term, dredging can virtually eliminate upriver sediment releases of PCBs.
- A recent Scenic Hudson national study of 89 river cleanup projects found dredging was preferred 90 percent of the time. Dredging reduced PCB levels in rivers and fish in locations such as the Fox River (Wisconsin), Manistique Harbor (Michigan), Cumberland Bay (New York) and Waukegan Harbor (Illinois).
- Dredge operations at rivers nationwide were minimally disruptive to lifestyle and recreation.
- River ecosystems will not be devastated and will quickly re-establish in a clean and healthy environment.

The EPA's PCB-removal plan combines plan-site source control with removing 100,000 pounds of PCBs from the river. Dredging will reduce cancer and noncancer dangers by up to 90 percent compared with just stopping contamination from GE's old plants.

See also Childhood Asthma and the Environment; Citizen Monitoring of Environmental Decisions; Cumulative Emissions, Impacts, and Risks

Web Resources

The Center for Environmental International Law. WTO "Supremacy Clause" in the POPs Convention, working paper. Available at www.ciel.org/Publications/pops2.html. Accessed January 21, 2008.

International Indian Treaty Council. Indigenous Environmental Network, press release. Available at www.treatycouncil.org/new_page_5213.htm. Accessed January 21, 2008.

Further Reading: Bargagli, Roberto. 2005. *Antarctic Ecosystems: Environmental Contamination, Climate Change, and Human Impact.* New York: Springer; Downie, David. 2002. *Northern Lights against POPs: Toxic Threats in the Arctic.* Montreal: McGill-Queen's Press; Harrad, Stuart. 2001. *Persistent Organic Pollutants: Environmental Behaviour and Pathways of Human Exposure.* New York: Springer; Johansen, Bruce Elliott. 2003. *The Dirty Dozen: Toxic Chemicals and the Earth's Future.* Westport, CT: Praeger; Johnston, Paul M., and Ruth Stringer. 2001. *Chlorine and the Environment: An Overview of the Chlorine Industry.* New York: Springer.

PESTICIDES

Chemicals used to kill insects, fungus, rats, and weeds are called *pesticides*. They can enter ecosystems and create damage. They can bioaccumulate up food chains and affect humans. Their widespread use makes environmentalists and communities uneasy despite improvements in public health due to their use. Some pesticide manufacturers label their products in confusing ways, which creates distrust. Some chemical manufacturers claim trade secrets when registering their pesticides with the EPA. Agribusiness points to high levels of productivity with their use. Many retail pesticides are sold every day to households. While pesticides are everywhere, many are concerned about exposure and health risks from them.

Many people are concerned about multiple exposures to pesticides. When those concerned are dismissed as hysterical housewives, the uneducated public, or extremist environmentalists, the seeds for controversy are sown. These are very serious concerns that demand explanation, and inadequate responses from government and industry do little to alleviate these concerns. Food and drinking water are sometimes contaminated from the same agricultural runoff. Some of the same pesticides used in industrial agriculture are also used in homes, hospitals, churches, schools, and day care centers. These are also places where vulnerable populations of the young, pregnant, and old can be more exposed. The human health effects from pesticide exposures are large. The large numbers of potentially affected people and the financial and social costs of exposure have not been considered in the formation of environmental policy around pesticides.

Questions about who is exposed to how much become questions about how much is safe for whom. There is a high level of concern around cumulative impacts and vulnerable populations that drives this controversy. Every niche of this controversy is laden with debating scientists, successful and unsuccessful lawsuits, agonizing government regulation and enforcement, and victims who will fight this issue to their literal death. The outline of these many battlegrounds shows the ferocity of this particular controversy. Some of the basic parameters of the current raging controversy about pesticide exposure of children are described by the Natural Resources Defense Council, a national U.S. environmental group. The summary findings of their research conclude:

- All children are disproportionately exposed to pesticides, compared with adults, due to their greater intake of food, water, and air per unit of body weight, their greater activity levels, narrower dietary choices, crawling, and hand-to-mouth behavior.
- Fetuses, infants, and children are particularly susceptible to pesticides compared with adults because their bodies cannot efficiently detoxify and eliminate chemicals, their organs are still growing and developing, and they have a longer lifetime to develop health complications after an exposure.
- Pesticides can have numerous serious health effects, ranging from acute poisoning to cancers, neurological effects, and effects on reproduction and development.
- Many pesticides that are never used indoors are tracked into the home and accumulate there at concentrations up to 100 times higher than outdoor levels.
- In nonagricultural urban or suburban households, an average of 12 different pesticides per home have been measured in carpet dust and an average of 11 different pesticide residues per household have been measured in indoor air in homes where pesticides are used.
- In an early 1990s nationwide survey of pesticide residues in urine in the general population, metabolites of two organophosphate pesticides, chlorpyrifos and parathion, were detected in 82 percent and 41 percent, respectively, of the people tested.

- In a rural community, all 197 children tested had urinary residues of the cancer-causing pesticide pentachlorophenol, all except six of the children had residues of the suspected carcinogen p-dichlorobenzene, and 20 percent had residues of the normally short-lived outdoor herbicide 2,4-D, which has been associated with non-Hodgkins lymphoma.

PESTICIDES IN AGRICULTURAL AREAS

The Natural Resources Defense Council did a special study of agricultural children and their exposure to pesticides, called *Trouble on the Farm: Growing Up with Pesticides in Agricultural Communities* (October 1998). Following is a summary of their conclusions.

- Children living in farming areas or whose parents work in agriculture are exposed to pesticides to a greater degree, and from more sources, than other children.
- The outdoor herbicide atrazine was detected inside all the houses of Iowa farm families sampled in a small study during the application season, and in only 4 percent of 362 nonfarm homes.
- Neurotoxic organophosphate pesticides have been detected on the hands of farm children at levels that could result in exposures above U.S. EPA designated safe levels.
- Metabolites of organophosphate pesticides used only in agriculture were detectable in the urine of two out of every three children of agricultural workers and in four out of every ten children who simply live in an agricultural region.
- On farms, children as young as 10 can work legally, and younger children frequently work illegally or accompany their parents to the fields due to economic necessity and a lack of child-care options. These practices can result in acute poisonings and deaths. (http://www.nrdc.org/health/kids/farm/exec.asp, May 15, 2007)

KINDS OF PESTICIDES

There are many different kinds of pesticides in use today. Pesticides are referred to according to the type of pest they control.

Chemical Pesticides

Some examples of chemical pesticides follow. Other examples are available in sources such as *Recognition and Management of Pesticide Poisonings*.

Organophosphate Pesticides

These pesticides affect the nervous system by disrupting the enzyme that regulates a chemical in the brain called acetylcholine. Most organophosphates are used as insecticides. Some are very poisonous. Both manufacturers of pesticides and government claim they are not persistent in the environment. Others claim they could be responsible for endocrine disruption in humans, as well as other nervous system impacts.

THE LANGUAGE OF PESTICIDES

Whether a given pesticide is safe partially depends on the application. Pesticides are application specific, and changes from these applications may pose hazards to people and the environment. While many of these terms have complex legal meanings, a working knowledge of the basic terms gives depth to this controversy. These are basic terms in the pesticide literature and can be found in the references and Web resources at the end of this entry. Pesticides that are related because they address the same type of pests include:

Algicides: Control algae in lakes, canals, swimming pools, water tanks, and other sites.

Antifouling agents: Kill or repel organisms that attach to underwater surfaces, such as boat bottoms.

Antimicrobials: Kill microorganisms (such as bacteria and viruses).

Attractants: Attract pests (for example, to lure an insect or rodent to a trap). (However, food is not considered a pesticide when used as an attractant.)

Biocides: Kill microorganisms.

Biopesticides: Certain types of pesticides derived from such natural materials as animals, plants, bacteria, and certain minerals.

Disinfectants and sanitizers: Kill or inactivate disease-producing microorganisms on inanimate objects.

Fumigants: Produce gas or vapor intended to destroy pests in buildings or soil.

Fungicides: Kill fungi (including blights, mildews, molds, and rusts).

Herbicides: Kill weeds and other plants that grow where they are not wanted.

Insecticides: Kill insects and other arthropods.

Microbial pesticides: Microorganisms that kill, inhibit, or outcompete pests, including insects or other microorganisms.

Miticides (also called acaricides): Kill mites that feed on plants and animals.

Molluscicides: Kill snails and slugs.

Nematicides: Kill nematodes (microscopic, worm-like organisms that feed on plant roots).

Ovicides: Kill eggs of insects and mites.

Pheromones: Biochemicals used to disrupt the mating behavior of insects.

Repellents: Repel pests, including insects (such as mosquitoes) and birds.

Rodenticides: Control mice and other rodents.

The term pesticide also includes these substances:

Defoliants: Cause leaves or other foliage to drop from a plant, usually to facilitate harvest.

Desiccants: Promote drying of living tissues, such as unwanted plant tops.

Insect growth regulators: Disrupt the molting, maturity from pupal stage to adult, or other life processes of insects.

Plant growth regulators: Substances (excluding fertilizers or other plant nutrients) that alter the expected growth, flowering, or reproduction rate of plants.

PESTICIDES AND PUBLIC HEALTH PROTECTION

Pesticides have had a strong historical role in limiting threats to the public health. Many states and localities use them for this purpose. Pests that are public health threats are eradicated or contained by the application of pesticides. When the government is applying the pesticides, some members of the resident community may object. Some people simply want to be free from any chemical intrusion and want that choice respected. Controversy can flare up at this point with public pest-eradication programs reliant on widespread application of persistent pesticides. Some members of the public want such a program to eliminate pestilence. Exactly what is a pest for these purposes?

The battleground for this type of pesticide controversy will often be the state agencies with responsibilities that include the use of pesticides. There are often the transportation departments that spray the side of the road with herbicides to keep weeds down. Weeds along a road can create a fire hazard in the hot, dry summer months of the United States.

WHEN IS A PEST A PUBLIC PEST?

Protecting public health goes beyond the general mandate to ensure the required safety of pesticides sold on the market for pest control. The Federal Insecticide, Fungicide, and Rodenticide Act (FIFRA) requires the U.S. Environmental Protection Agency (EPA), in coordination with the U.S. Department of Health and Human Services (HHS) and U.S. Department of Agriculture (USDA), to identify pests of significant public health importance and, in coordination with the Public Health Service, to develop and implement programs to improve and facilitate the safe and necessary use of chemical, biological, and other methods to combat and control such pests. FIFRA defines the term pest as meaning:

1. any insect, rodent, nematode, fungus, weed, or
2. any other form of terrestrial or aquatic plant or animal life or virus, bacteria, or other micro-organism (except viruses, bacteria, or other micro-organism on or in living man or other living animals) that the Administrator declares to be a pest under section 25(c)(1).

The EPA has broadly declared the term pest to cover each of the organisms mentioned except for the organisms specifically excluded by the definition. Following is a brief description of the identified pests or categories of pests and an explanation for designating each as a public health pest.

- Cockroaches. Cockroaches are controlled to halt the spread of asthma, allergies, and food contamination.
- Body, head, and crab lice. These lice are surveyed for and controlled to prevent the spread of skin irritation and rashes, and to prevent the occurrence of louse-borne diseases such as epidemic typhus, trench fever, and epidemic relapsing fever in the United States.
- Mosquitoes. Mosquitoes are controlled to prevent the spread of mosquito-borne diseases such as malaria; St. Louis, Eastern, Western, West Nile, and LaCrosse encephalitis; yellow fever; and dengue fever.

- Various rats and mice. The listed rats and mice include those that are controlled to prevent the spread of rodent-borne diseases and contamination of food for human consumption.
- Various microorganisms, including bacteria, viruses, and protozoans. The listed microorganisms are the subject of control programs by public health agencies and hospitals for the purpose of preventing the spread of numerous diseases.
- Reptiles and birds. The listed organisms are controlled to prevent the spread of disease and the prevention of direct injury.
- Various mammals. The listed organisms have the potential for direct human injury and can act as disease reservoirs (i.e., rabies, etc.).

It is possible that this list may need to be changed. Should any additional species be found to present public health problems, the EPA may determine that it should consider them to be pests of significant public health importance under FIFRA. The EPA is supposed to update the list of pests of significant public health importance.

PESTICIDES AND FOOD QUALITY

The Food Quality Protection Act placed requirements on the EPA related to public health and pesticides. The EPA considers risks and benefits of pesticides that may have public health uses. The EPA regulates certain pesticides that might be found in drinking water by setting maximum contaminant limits.

PESTICIDE SPRAY DRIFT: ANOTHER CASE OF INVOLUNTARY EXPOSURE

Another controversial risk to public health is from pesticide spray drift. The EPA defines pesticide spray drift as the physical movement of a pesticide through air at the time of application or soon thereafter, to any site other than that intended for application (often referred to as off-target). This can affect the health of neighboring communities and farms, especially organic farms.

Pesticide drift can affect human health and the environment. Spray drift can result in pesticide exposures to farmworkers, children playing outside, and the ecosystem. Drift can also contaminate a home garden or another farmer's crops.

There are many reported complaints of pesticide spray drift each year. Reports of exposures of people, plants, and animals to pesticides due to off-target drift are called drift incidents. They are part of an important component in the scientific evaluation and regulation of the uses of pesticides. The EPA is supposed to consider all of these routes of exposure in regulating the use of pesticides. A major criticism of the EPA approach is that it does not measure the cumulative effects of pesticide exposure. Another battleground is the weak enforcement of this environmental protection policy. EPA policy relies on complaints of drift incidents and on labeling. This is often seen as a weak and ineffective response by many who are subject to repeated drift incidents. If no people are nearby, the environmental impacts of drift incidents may accrue over years, eventually working their way into land and water systems. This underscores the necessity of a cumulative and ecosystem risk analysis, which are battlegrounds themselves. The EPA allows some degree of drift of pesticide particles in almost all applications. It assumes pesticide applications are made in responsible ways by trained operators. This is not the case in many instances. In making their decisions about pesticide applications, prudent and responsible applicators must consider all factors, including wind speed, direction, and other weather conditions; application equipment; the proximity of people and sensitive areas; and product label directions. A prudent and responsible applicator must refrain from application under conditions that can cause pesticide drift. They decide whether or not to apply a pesticide. It is their responsibility to know and understand a product's use restrictions, but most do not. The practical result is potential human health effects from chemical overexposure.

The EPA conducts ecological risk assessments to determine what risks are posed by a pesticide and whether changes to the use or proposed use are necessary to protect the environment. Many plant and wildlife species can be found near or in cities, agricultural fields, and recreational areas. Before allowing a pesticide product to be sold on the market, the EPA ensures that the pesticide will not pose any unreasonable risks to wildlife and the environment. They do this by evaluating data submitted in support of registration regarding the potential hazard that a pesticide may pose to nontarget fish and wildlife species.

PESTICIDES, WATER QUALITY, AND SYNERGY

When pesticides are applied on fields, gardens, parks, and lawns, some of the chemicals run off the treated site. More than 80 percent of urban streams and more than 50 percent of agricultural streams have concentrations in water of at least one pesticide, mostly those in use during the study period, that exceed a water-quality benchmark for aquatic life. Water-quality benchmarks, set by the EPA, are estimates of pesticide concentrations that the agency says may have adverse effects on human health, aquatic life, or fish-eating wildlife. Insecticides, particularly diazinon, chlorpyrifos, and malathion, frequently exceed aquatic-life benchmarks in urban streams. Most urban uses of diazinon and chlorpyrifos, such as on lawns and gardens, have been phased out since 2001 because of an EPA-industry agreement. In agricultural streams, the pesticides chlorpyrifos, azinphos-methyl, p,p'-DDE, and alachlor are among those most often found above benchmarks. While the standard benchmarks were not exceeded for human health, recent studies and decades of incomplete risk assessments suggest that EPA benchmarks are severely underestimated. This is a very controversial scientific issue. If synergy and cumulative impacts increase public health risk, then new regulations with different standards could become law.

Pesticides seldom occur alone in the real world, but rather almost always as complex mixtures. Most stream samples and about half of well samples contain two or more pesticides, and frequently more. The potential effects of contaminant mixtures on people, aquatic life, and fish-eating wildlife are still poorly understood. Most toxicity information and the water-quality benchmarks are developed for individual chemicals. The common occurrence of pesticide mixtures, particularly in streams, means that the total combined toxicity of pesticides in water, sediment, and fish may be greater than that of any single pesticide compound that is present. Studies of the effects of mixtures are still in the early stages, and it may take years for researchers to attain major advances in understanding the actual potential for effects. A recent study by researchers at the University of California–Berkeley finds that pesticide mixtures harm frogs at levels that do not produce the same effects alone, often levels 10 to 100 times below EPA standards. This has implications for local governments and other water providers. Drinking water providers are faced with a dilemma about how to deal with the twin problem of killing dangerous bacteria while not increasing the chemical health risks for pregnant women and healthy infants. Pesticides are getting into the drinking water sources for millions of people in the United States. These contaminants combine with disinfectants, such as chlorine (added by drinking water providers to kill dangerous viruses, bacteria, and pathogens), forming disinfectant by-products that are associated with increases in birth defects and miscarriages. This drives the concern that the pesticides in our drinking water cannot be addressed by the chemical-by-chemical regulatory approach of government.

PESTICIDES: THE FIGHT IN CONGRESS

Pesticide regulation and control are very controversial in Congress. Presidential administrations also affect this controversy in different ways. One very controversial issue is whether pesticides should be tested on people. Most scientific studies generalize from mice or rats to humans. Some are concerned that this species generalization may be inaccurate, especially if the effects of other chemicals in the environment are considered. As the EPA and other governmental agencies struggle to develop standards and regulations around pesticides, the need for exact data increases. As environmentalists and toxic tort lawyers challenge industry in the courts, the need for data that support human injury also increases. However, most individuals, when given a free choice, choose not be tested for pesticide safety. A core question is whether safe standards for humans can be developed without testing pesticides on humans. Scientifically this may be difficult, but it is demanded by the public.

On June 29, 2005, one day after the George W. Bush administration's proposal to allow industry to test pesticides on people was leaked to the media, the Senate acted to block the testing of pesticides on humans. By a vote of 60–37, the Senate adopted a bipartisan amendment to the 2006 Interior and Environment appropriations bill, which blocks the Environmental Protection Agency for one year from spending tax dollars to fund or review studies that intentionally expose people to pesticides. The U.S. Senate also voted 57–40 to adopt another amendment that requires the EPA to review the ethical ramifications of human pesticide testing.

RESEARCH NEEDED

While the effects of pesticides increase there is a continued need for more research. Pesticide controversies are tied to cumulative risk controversies and therefore will not decrease. Given recent scientific evidence of pesticide synergy, it is likely that public policy development will require more data and information to react and to plan for pesticide public health issues. This is an emerging battleground that pits public health interests against chemical manufacturers of pesticides. It will be difficult for scientists to exercise independent professional judgment given the disparity of power and resources between the two interests.

PESTICIDES: INERT INGREDIENTS IN PESTICIDE PRODUCTS— ABOUT TOLERANCE REASSESSMENT

Pesticide Tolerances

How much of a chemical an individual can take before an adverse response occurs is called a *tolerance limit,* or a *dose response.* In investigating the safety of pesticides researchers determine the tolerance limit for an average person. This model person was usually a 155-lb. white male, among the healthiest demographic groups in U.S. society. There are two controversial problems with this method. There is a large dose-response variance in the U.S. population. Something as simple as aspirin can have a 100-fold dose-response difference. This model does not take vulnerability into account and thus underestimates actual risk from pesticides for many parts of the U.S. population. The second problem is that the EPA approach does not take into consideration the accumulated pesticide impacts on a given person. From the human health perspective this is very important. One dose of a chemical may not kill you, but 10,000 small exposures could. The difference to the human body or to the environment if the exposure comes from one bad episode of pesticide drift or 10,000 exposures to residential lawn pesticides is small.

The EPA sets limits on the amount of pesticides that may remain in or on foods. These limits are called tolerances. The tolerances are set based on a risk assessment and are enforced by the Food and Drug Administration. More information on tolerances is provided at Pesticide Tolerances (www.epa.gov/opprd001/inerts/tol.htm).

The EPA is reassessing tolerances (chemical-residue limits in food) and exemptions from tolerances for inert ingredients in pesticide products to ensure that they meet the safety standard established by the Food Quality Protection Act (FQPA) in 1996. FQPA requires the reassessment of inert ingredient tolerances and tolerance exemptions that were in place prior to August 3, 1996. Since the passage of FQPA, EPA has been reassessing inert ingredients and will complete the remaining assessments by August 2006. Completed inert ingredient tolerance reassessment decision documents are now available on the Tolerance Reassessment Status page (www.epa.gov/pesticides/tolerance/reassessment.htm). Refer to the U.S. Government Printing Office–issued Code of Federal Regulations (CFR) for complete and current information on inert (other) ingredient tolerances and tolerance exemptions. The CFR describes how each inert ingredient may be used.

POTENTIAL FOR FUTURE CONTROVERSY

Pesticide controversies will continue to dominate environmental policy in the courts, legislatures, agencies, and communities. They affect private households, local and state governments, big and small industries, agribusiness, the public health of communities, and many who rely on them. They have large impacts on the environment that may accumulate and bioaccumulate. Many do pose risks to vulnerable populations, and many more could if they mix with other chemicals in the real world. Who is to bear the risk of unknown harm, whether it is from pesticides or a lack thereof? Science will have a big role to play in this controversy, and there are politically powerful forces in opposition to each other about this controversy.

See also Cumulative Emissions, Impacts, and Risks; Ecosystem Risk Assessment; Genetically Modified Food; Organic Farming

Web Resources

National Pesticide Information Center. Available at npic.orst.edu/. Accessed January 21, 2008.
Northwest Coalition for Alternatives to Pesticides. Available at www.pesticide.org/. Accessed January 21, 2008.
U.S. Environmental Protection Agency. Pesticides: Health and Safety. Recognition and Management of Pesticide Poisonings. Available at www.epa.gov/pesticides/safety/health-care/handbook/handbook.htm. Accessed January 21, 2008.
U.S. Environmental Protection Agency. Pesticides: Science and Policy. Committee to Advise on Reassessment and Transition—Paper #5. Available at www.epa.gov/pesticides/carat/2000/june/paper5.htm. Accessed January 21, 2008.

Further Reading: Beres, Samantha. 2002. *Pesticides: Critical Thinking about Environmental Issues.* Farmington Hills, MI: Greenhaven Press; Jacobs, Miriam, and Barbara Dinham, eds. 1992. *Silent Invaders: Pesticides, Livelihoods and Women's Health.* London: Zed Books; Matthews, Graham A. 2006. *Pesticides: Health, Safety and the Environment.* Oxford: Blackwell Publishing; Nash, Linda Lorraine. 2007. *Inescapable Ecologies: A History of Environment, Disease, and Knowledge.* Berkeley: University of California Press; National Research Council. 2000. *The Future Role of Pesticides in U.S. Agriculture.* Washington, DC: National Academies Press; Wargo, John. 1998. *Our Children's Toxic Legacy: How Science and Law Fail to Protect Us from Pesticides.* New Haven, CT: Yale University Press.

POLLUTION RIGHTS OR EMISSIONS TRADING

Emissions trading is a regulatory environmental policy to reduce the cost of pollution control by providing economic incentives to regulated industries for achieving reductions in the emissions of pollutants. A central authority, such as an air pollution control district or a government agency, sets limits or caps on each regulated pollutant. Industries that intend to exceed their permitted limits may buy emissions credits from entities that are able to stay below their permitted limits. This transfer is normally referred to as a trade. This is a new policy in the United States. Questions that may become controversies include whether all emissions are measured. Another more basic question is whether society can still allow polluters to buy their way out of responsibility for environmental and community impacts.

HARNESSING MARKET FORCES FOR A SAFER ENVIRONMENT?

Market-based environmental policies for reducing pollution include many economic or market-oriented incentives. These include tax credits, emissions fees, or emissions trading. There are many types of emissions-trading approaches; the one used by EPA's Clean Air Market Programs is called allowance trading or cap and trade and has the following key features:

1. An emissions cap: a limit on the total amount of pollution that can be emitted (released) from all regulated sources (e.g., power plants); the cap is set lower than historical emissions to cause reductions in emissions.
2. Allowances: an allowance is an authorization to emit a fixed amount of a pollutant
3. Measurement: accurate tracking of all emissions
4. Flexibility: sources can choose how to reduce emissions, including whether to buy additional allowances from other sources that reduce emissions
5. Allowance trading: sources can buy or sell allowances on the open market
6. Compliance: at the end of each compliance period, each source must own at least as many allowances as its emissions

WHAT IS THE U.S. ENVIRONMENTAL POLICY ON EMISSIONS TRADING?

According to the U.S. Environmental Protection Agency (EPA), cap and trade is a policy approach to controlling large amounts of emissions from a group of sources at a cost that is lower than if sources were regulated individually. The approach first sets an overall cap, or maximum amount of emissions per compliance period, that will achieve the desired environmental effects. Permits to emit are then allocated to pollution sources, and the total number of allowances cannot exceed the cap. The main requirement is that pollution sources completely and accurately measure and report all emissions. There is grave concern about this premise. Since not all emissions are counted now, many are concerned that this lack of specific reporting will only hide pollution.

WHERE HAS THIS APPROACH BEEN USED SUCCESSFULLY?

Cap and trade was first tried in the United States to control emissions that were causing severe acid rain problems over very large areas of the country.

Legislation was passed in 1990 and the first compliance period was 1995. Sulfur dioxide (SO_2) emissions have fallen significantly, and costs have been even lower than the designers of the program expected. The U.S. Acid Rain Program has achieved greater emissions reductions in such a short time than any other single program to control air pollution. A cap and trade program also is being used to control SO_2 and nitrogen oxides (NO_x) in the Los Angeles, California, area. The Regional Clean Air Incentives Market (RECLAIM) program began in 1994.

The regulating agency (e.g., EPA) must:

- be able to receive the large amount of emissions and allowance transfer data and assure the quality of those data
- be able to determine compliance fairly and accurately
- strongly and consistently enforce the rule

Allowance trading is the centerpiece of EPA's Acid Rain Program, and allowances are the currency with which compliance with the SO_2 emissions requirements is achieved. Through the market-based allowance trading system, utilities regulated under the program, rather than a governing agency, decide the most cost-effective way to use available resources to comply with the acid rain requirements of the Clean Air Act. Utilities can reduce emissions by employing energy conservation measures, increasing reliance on renewable energy, reducing usage, employing pollution-control technologies, switching to lower-sulfur fuel, or developing other alternate strategies. Units that reduce their emissions below the number of allowances they hold may trade allowances with other units in their system, sell them to other utilities on the open market or through EPA auctions, or bank them to cover emissions in future years. Allowance trading provides incentives for energy conservation and technology innovation that can both lower the cost of compliance and yield pollution-prevention benefits, although this is controversial.

The Acid Rain Program established a precedent for solving other environmental problems in a way that minimizes the costs to society and promotes new technologies.

WHAT ARE ALLOWANCES?

An allowance authorizes a unit within a utility or industrial source to emit one ton of SO_2 during a given year or any year thereafter. At the end of each year, the unit must hold an amount of allowances at least equal to its annual emissions, that is, a unit that emits 5,000 tons of SO_2 must hold at least 5,000 allowances that are usable in that year. However, regardless of how many allowances a unit holds, it is never entitled to exceed the limits set under Title I of the act to protect public health. Allowances are fully marketable commodities. Once allocated, allowances may be bought, sold, traded, or banked for future use. Allowances may not be used for compliance prior to the calendar year for which they are allocated.

WHO MAY PARTICIPATE IN ALLOWANCE TRADING?

Allowances may be bought, sold, and traded by any individual, corporation, or governing body, including brokers, municipalities, environmental groups, and private citizens. The primary participants in allowance trading are officials designated and authorized to represent the owners and operators of electric utility plants that emit SO_2.

HOW WOULD COMPLIANCE BE DETERMINED?

At the end of the year, units must hold in their compliance subaccounts a quantity of allowances equal to or greater than the amount of SO_2 emitted during that year. To cover their emissions for the previous year, units must finalize allowance transactions and submit them to the EPA by March 1 to be recorded in their unit accounts. If the unit's emissions do not exceed its allowances, the remaining allowances are carried forward, or banked, into the next year's account. If a facility's emissions exceed its allowances, it must pay a penalty and surrender allowances for the following year to the EPA as excess emission offsets.

Emissions trading or marketable rights have been in use in the United States since the mid-1970s. The advocates of free-market environmentalism sometimes use emissions trading or marketable rights systems as examples to support the theory that free markets can handle environmental problems.

The idea is that a central authority will grant an allowance to entities based on a measure of their need or their previous pollution history. For example an allowance for greenhouse gas emissions to a country might be based on total population of the country or on existing emissions of the country. An industrial facility might be granted a license for its current actual emissions. If a given country or facility does not need all of its allowance, it may offer it for sale to another organization that has insufficient allowances for its emission production. Environmentalists point out that this only increases environmental impacts to the carrying capacity, and beyond, of the environment. They observe that industry is supposed to reduce its emissions to the greatest extent possible under current environmental law. Claims that emissions will somehow be reduced now, or at least shifted to where it could saturate another environment, are not viewed as credible. This lays the foundation for controversy. Communities point out that the cumulative impact of already-existing industries is a concern and possible public health risk.

PROMINENT TRADING SYSTEMS

The most common policy example of an environmental emissions trading system is the sulfur dioxide trading system contained in the Acid Rain Program of the 1990 Clean Air Act. Under the program, sulfur dioxide emissions are to be reduced by 50 percent from 1980 to 2010. In 1997, the state of Illinois adopted

POLLUTION OR EMISSIONS RIGHTS TRADING

Interactive Online Reports of Emissions Markets

The market for emissions rights depends on skilled and knowledgeable entrepreneurs taking risks. As this is a new policy, getting information out to entrepreneurs is a high policy priority. To accommodate this need, interactive online reports of emissions markets are available.

For each report listed here, you can:

- Search for information based on certain criteria, as described here.
- Download records from a report by clicking the Download Records button.
- They will be saved in comma-delimited format (commas between fields).

Account Report	SO_2 | NO_x
Allowances Held Report	SO_2 | NO_x
Serial Number Report	SO_2 | NO_x
Transactions	SO_2 | NO_x
Account Owners Report	SO_2 | NO_x

These data are updated daily, between 5:00 A.M. and 6:00 A.M., Eastern Time.
Allowances held by all accounts.
SO_2 (includes trades through February 24, 2006; approx. 665 Kb).

The SO_2 allowance database is too large to download using the allowances held report, so this information is offered in a single compressed file containing the following information for every SO_2 allowance account:

- account number
- account name
- allowance use year
- quantity of allowances held for the use year
- starting and ending serial number

NO_x

The NO_x database is small enough to be downloaded in its entirety using the NO_x allowances held report (leave all entries blank and submit the form). Although not offered as a single compressed file, the advantage is that the information is updated daily instead of weekly.

Report Descriptions
Search Criteria
Detailed description (i.e., metadata) for ATS and NATS

Table P.2 Report Descriptions

Report Name	Data Included in Report
Account Report Create report: SO2	NOx	Account ID, Account/Plant Name, State (where the plant is located, if applicable), Representative ID, Representative Name, Alternate ID, Alternate Name, Firm Represented by Representative, Representative Contact Info (Address, Phone, and Fax)
Allowances Held Report Create report: SO2	NOx	Account ID, Account Plant Name, Allowance ("vintage") Year, Number of Allowances Held (of the associated vintage), State (where the plant is located, if applicable), and Representative Name.
Serial Number Report Create report: SO2	NOx	Lists Account ID, Account Plant Name, Allowance ("vintage") Year, Starting Serial Number, Ending Serial Number, the Number of Allowances Represented in the "Block," State (where the plant is located, if applicable), Representative Name and Status of the Allowance Block (this flag indicates allowances that will expire; these are allowances allocated, under Early Reduction authority, to NO_x Budget Program sources).
Transactions Report Create report: SO2	NOx	Lists Transaction Number, Transaction Type, Transferee/Transferor Account ID, Transferee/Transferor Account/Plant Name, Transferee/Transferor State (where the plant is located, if applicable), Transferee/Transferor Representative Name, Total Number of Allowances Involved in the Transaction, Date the Transaction Was Confirmed (recorded in the Allowance Tracking System); Allowance Year ("vintage"), Start and End Serial Numbers (of an allowance "block" involved in the transaction), and the Number of Allowances Represented in the "Block" (defined by the Start and End Serial Numbers).
Account Owner Report Create report: SO2	NOx	Lists Account ID, Account/Plant Name, State (where plant is located, if applicable), Owner/Binding Party Name, and Representative Name.

a trading program for volatile organic compounds in the Chicago area, called the Emissions Reduction Market System. Beginning in 2000, over 100 major sources of pollution in eight Illinois counties began trading pollution credits. In 2003, New York State proposed and attained commitments from nine northeastern states to cap and trade carbon dioxide emissions. States and regions of the United States are pursuing more of these policies.

The European Union Greenhouse Gas Emission Trading Scheme is the largest multinational, greenhouse gas emissions trading scheme in the world. It started in January 2005 and all 25 member states of the European Union participate in it.

POTENTIAL FOR FUTURE CONTROVERSY

Emissions trading is a new and experimental policy in the United States. However, it exposes major flaws in the current U.S. environmental regulatory regime. Most environmental information about some of the biggest and unknown environmental impacts comes from self-reporting and may not be accurate. Many industries self-report whether they emit enough to even require any type of permit or oversight. Once regulated, emissions are simply permitted, and amounts are self-reported by industry. Emissions trading may be seen as a free-market band-aid to a young, weak, and incomplete public policy of environmental protection. It opens up large holes in the current system and may inflame environmentalists and the public depending on how it is implemented. In a society moving toward concepts like sustainability, emissions trading may be challenged in its present form.

See also Cumulative Emissions, Impacts, and Risks; Environmental Audits and Environmental Audit Privileges; Sustainability

Web Resources

Environmental Economics. Problems of Emissions Trading. Available at www.env-econ. net/2006/12/problems_with_e.html. Accessed January 21, 2008.

International Emissions Trading Association. Available at www.ieta.org/ieta/www/pages/ index.php. Accessed January 21, 2008.

Further Reading: Grubb Michael, and Karsten Neuhoff. 2006. *Allocations, Incentives and Industrial Competitiveness under the EU Emissions Trading Scheme*. London: Earthscan; Hansjürgens, Bernd. 2005. *Emissions Trading for Climate Policy: US and European Perspectives*. Cambridge: Cambridge University Press; de Jong, Cyriel, and Kasper Walet, eds. 2004. *A Guide to Emissions Trading: Risk Management and Business Implications*. London: Risk Books.

POVERTY AND ENVIRONMENT IN THE UNITED STATES

Residents of environmentally degraded neighborhoods are generally poorer, with unemployment rates nearly 20 percent higher than the national average. Some argue for no government intervention in this instance saying that the market will balance out over time. Others argue for strong environmental intervention in these neighborhoods saying that cumulative impacts and a concern for the future should be the driving forces of strong governmental intervention in environmental protection.

Poverty and environmental degradation often accompany each other in the United States. For some statistical reasons, many scientists consider it very difficult to disaggregate race and income data when searching for cause and effect. Poverty itself is a policy term of art. It can mean different things under different policies, but basically it means that income is so low it is difficult to meet basic food, shelter, and clothing needs. Generally, 5–15 percent of the U.S. population are relatively poor. They are more often unemployed and live in communities desperate for jobs. This often makes community leaders accept environmentally undesirable economic development activities such as waste sites and waste transfer stations. Concentrations of metal fabrication plants, slaughterhouses, tanneries, incinerators, and auto body repair facilities are other generally undesirable land uses associated with poor communities. The lower one's income, the higher the probability of exposure to environmental and occupational hazards at home and on the job. Also, the lower one's income, generally, the greater the risk of diseases caused or exacerbated by environmental factors. Food, especially organic food, is expensive for poor people. Transportation is very difficult for poor people. This makes it difficult for the poor to find employment, housing, and food. Poor neighborhoods generally receive fewer municipal services, like parks and open outdoor recreation areas.

JOB BLACKMAIL

Many communities rely on one or more industries as their main economic base. Some of these communities may have begun as company towns. Many of these communities bear the brunt of industrial pollution. However, the threat of losing a job and home due to labor downsizing allegedly caused by environmental regulations is a controversial one. Employers and industry supporters use the threat of job loss to blackmail or force workers and communities into accepting unhealthy workplaces and community environmental damage.

For example, in a 1995 job blackmail episode, the Raytheon Corporation, a large employer, threatened to move jobs from Massachusetts unless the state rewrote its corporate income tax code for defense contractors, to save the company about three-fourths of its tax bill. That facility and industry still enjoyed the benefits and privileges of the services provided by the state, they just paid less for them. Then the state's manufacturers association demanded the same tax loophole for all defense manufacturers. Then Fidelity Investments demanded the same tax loophole for mutual fund companies.

The effects of environmental protection and regulation on labor demand are small compared to other factors. They are found to produce a slight employment benefit and economic stimulus due to the effects of public expenditure on environmental protection, corporate investment in abatement measures and accelerated technological innovation, the stimulus to the environmental protection industry, and the adoption of more labor-intensive production processes.

Environmental regulation may be responsible for some plant closures and layoffs, but less than 0.1 percent of layoffs annually. This is a hotly disputed number, and there are many other factors. Many plant closures and layoffs are caused by regular business processes such as business failures, external competition, technological change, corporate restructuring, relocation, product changes, and seasonal variations in activity. The number of people laid off was also less than the extra jobs created by environmental protection. There are few examples of relocation of U.S. plants to pollution havens. In general, environmentally induced capital flight has not occurred, because the cost of environmental regulation is small compared to overall business, costs are only one factor in relocation decisions, and modern production technology incorporates pollution-control devices for most regulated air and water pollutants. Government and industry-funded macroeconomic modeling of the economic and employment effects of environmental protection overestimates the adverse impacts. In 13 cases of new national environmental laws, the models had overestimated compliance costs by up to 2,900 percent, and in one case had predicted compliance costs where there were none. Such models had also predicted that proposed changes to the Clean Air Act in the early 1990s would cause a major economic downturn and from 200,000 to 2,000,000 job losses. The recession did not materialize, and fewer than 7,000 workers required assistance because their jobs were affected by the new laws. The models also usually inflated the benefits of environmentally contentious developments. Such models are inaccurate because they proceed from unrealistic assumptions that militate against environmental protection and in favor of permitting business as usual, including that the economy cannot respond innovatively and flexibly to environmental regulation so as to minimize costs and create business opportunities, and that governments will pursue the most costly and least flexible means of implementing such regulation. They also fail to take account of the direct and indirect economic, social, and environmental benefits of the regulation.

There is no strong case for a large overall employment gain from strong environmental regulation or an economy-wide shift to clean technology. The subsequent employment history and economic well-being of displaced workers varied according to the overall state of regional economies and the adequacy of federal government retraining and income-support policies. Timber workers found themselves in a strong regional economy that enabled most to be re-employed. Coal miners, however were not able to find jobs in their depressed regional economy. The usefulness of retraining and job-search programs was limited by the strict eligibility rules, time-limited nature, and stingy levels of U.S. unemployment benefits. U.S. workers are more susceptible to job-blackmail tactics by employers than European workers.

POVERTY

In the United States in 2008, 12.7 percent of the population lives below the poverty line. Poverty affects housing quality, educational opportunities, health, and employment opportunities. It also affects the quality of the air, water, and land. Air conditioning, filtered and bottled water, and clean soil all come at a price. Being poor means that your air, water, and land are more likely to be laden with environmental stressors. What are the tangible measures of poverty in the United States?

$35,000—basic-needs budget for a U.S. family of four (two adults, two children).

$19,157—poverty line for a family of four (two adults, two children) in the United States in 2004, as established by the U.S. Census Bureau.

12.7—percentage of U.S. citizens living below the poverty line in 2004 (37 million people).

8.6—percentage of non-Hispanic Caucasians living below the poverty level in 2004.

9.8—percentage of Asians living below the poverty level in 2004.

21.9—percentage of Hispanics living below the poverty level in 2004.

24.3—percentage of Native Americans living below the poverty level in 2004.

24.7—percentage of African Americans living below the poverty level in 2004.

The precautionary principle has flourished in international policy statements and agreements. It was first recognized in the World Charter for Nature, which was adopted by the UN General Assembly in 1982. Later it was adopted in the First International Conference on Protection of the North Sea in 1984. The UN Conference on Environment and Development in Rio de Janeiro in 1992 incorporated the precautionary principle in several important documents. UN leaders adopted Principle 15 and advocated the widespread international application of the precautionary principle. Principle 15 states that:

> In order to protect the environment, the precautionary approach shall be widely applied by states according to their capabilities. Where there are threats of serious or irreversible damage, lack of full scientific certainty shall not be used as a reason for postponing cost-effective measures to prevent environmental degradation.

The precautionary principle was also formally adopted by countries of the European Union in the Treaty of Maastricht in 1992. It has since been extended from environmental issues to developments related to human health.

U.S. APPLICATIONS

The minimal concept of precaution underpins some U.S. environmental policy, such as the requirement for environmental impact statements before starting major projects with significant environmental impacts using federal funds. Most current environmental laws in the United States focus on cleaning up and controlling damage rather than preventing it. Even environmental impact statement recommendations are basically advisory, and the law is riddled with categorical exceptions. These early U.S. environmental policies do not sufficiently protect people and the environment from irreparable damage. The Clean Air Act, the Clean Water Act, and other laws basically regulate pollution from large polluters but do not prevent it. As effects accumulate, population increases, and society realistically examines sustainability, precautionary principles will expand from international treaties to local U.S. land-use practices and even further. This brings much greater environmental scrutiny to industrial and other land-use practices that have environmental impacts that accumulate and/or affect the ecosystem. State and local governments are an emerging battleground for the precautionary principle. In these battlegrounds, the environmental scrutiny is often on cleanup of past environmental degradation. Liability for these cleanup costs will be enormous, and anything that brings environmental scrutiny to these past and present practices is strongly resisted by private property and industrial interests.

Most communities adopt a basic approach. The precautionary principle is common sense to them. Proponents argue that they need to prevent potentially dangerous practices at the beginning of a project rather than afterwards. Many communities have had difficult relationships with industrial residents and may feel that government environmental enforcement is not reliable or effective. Local leaders feel that they often do not know for sure what harm there will be until people have suffered or the damage is irreparable. They compare these uncertain risks with the promise of increased employment.

PUBLIC HEALTH BASIS OF THE PRECAUTIONARY PRINCIPLE IN LAND-USE PLANNING PROCESSES

Public health, environmental regulation, and land-use planning are separate battlegrounds and seldom relate to each other in the United States. Each has separate processes, funding streams, and forums. An international organization of local government has recognized the environmental and public health harms caused by land-use decisions. They have begun to advocate for the precautionary principle as a way of making environmentally sensitive land-use decisions that incorporate the public health. In September 2003, the National Association of County and City Health Officials (NACCHO) passed a resolution advocating for the adoption of the precautionary principle. The resolution prompted several U.S. municipalities to adopt the precautionary principle in their land-use planning processes, such as Portland, Oregon, and San Francisco, California. Land-use laws generally have some basis in protecting the public health, safety, and welfare. The resolution states:

Whereas, land use decisions may contribute to:

Health inequities; an increase in health and safety risks, poor quality housing, unstable neighborhoods, unsustainable ecosystems, and poor quality of life can be created; asthma mortality is approximately three times higher among African Americans than it is among whites; the elderly and people with disabilities are disproportionately affected by a lack of sidewalks and depressed curbs; and

Chronic disease; more than 25 percent of adults in the United States are obese, and more than 60 percent do not engage in enough physical activity to benefit their health; research has shown that a healthy diet and physical activity can prevent or delay type 2 diabetes; and

Increased traffic congestion, reliance on the automobile, and increased pedestrian and bicyclist vulnerability; commuting stress has increased in recent years, while there has also been a decline in social capital (community connectedness); one pedestrian is killed in a vehicle accident every 108 minutes and injured every 7 minutes; and

Decreased air quality and increased pollution emissions; motor vehicles are the largest source of manmade urban air pollution, and the EPA attributes 64,000 premature deaths per year to air pollution; between 1980 and the mid 1990s, the rate of people with asthma rose by 75 percent; and

Decreased water quality; according to the EPA, soil erosion, and destruction of wetlands threaten surface and ground water quality, which may be drinking and/or recreational water sources; runoff from point and non-point sources pollute waterways, and is exacerbated when the amount of impervious surface in an area is increased; and

Loss of greenspace and land conversion; greenspace provides benefits for air and water quality, as well as for the physical and mental health of people; sprawling development consumes 1.2 million acres of productive farmland per year; according to the American Farmland Trust, land is being developed at two times the population growth rate; and

Inappropriate hazardous materials facilities siting, transportation, and storage; exposure to heavy metals has been linked with certain cancers, kidney damage, and developmental retardation; and areas zoned for hazardous materials storage that contain toxic-waste facilities are often located near housing for poor, elderly, young, and minority residents.

While not all these justifications will apply to all areas, many of them do currently apply to U.S. cities. After calling for the precautionary principle, the NACCHO resolution suggests three ways to make it work:

1. integrate public health perspectives and practice (which are based on prevention) into land-use planning;
2. ensure early, sustained, and effective participation by affected community members in all stages of land-use and zoning decisions; and
3. dedicate more resources to getting public health people involved in land-use decisions through training, development of tools, technical assistance, and other support.

WHAT ACTION IS NECESSARY UNDER THE PRECAUTIONARY PRINCIPLE?

There are many stakeholders to this controversy. Land developers, construction companies, bankers, real estate corporations, industrial developers, and venture capital specialists strongly object to anything that can delay or derail a project with environmental impacts. Many times their objections and resistance form a battleground when they are required to do specific actions. Precautionary action may include a sustainability focus, restricting or phasing out practices or substances that irreparably harm the environment, developing new technologies with softer ecological footprints (i.e., using renewable sources of energy), or sometimes denying the project. If a project proponent comes forward with a development with significant environmental impacts, the possibility of the project being rejected can be substantial under the precautionary principle. The proponent may have to do expensive scientific studies that indicate their project should not be denied. These last two dynamics, denial of the project and required scientific study of all impacts, are new in the United States, and very controversial. Currently, when human and environmental damage occurs, victims have a hard time proving in court that a product or activity was responsible. The precautionary principle shifts the burden of proof to the alleged perpetrator. Project proponents must prove its safety and are held responsible for damages and for mitigation and remediation of environmental impacts.

REGULATORY CONCERNS

The precautionary principle requires regulatory action on the basis of possible and currently unknown unmanageable risks. Business and regulatory decisions are made to stop or slow development, on the basis of what we do not know, according to industry advocates. The environmentalists respond that in interpreting the precautionary principle, it involves a proportionality of response to ensure that the selected degree of restraint is not too expensive and that it does protect public health and the environment.

INDUSTRY'S RESPONSE TO THE PRECAUTIONARY PRINCIPLE

In the battle for public opinion and sentiment, industrial trade representatives and others develop public relations strategies. These are part of many corporate planning processes. In terms of environmental public relations, these strategies seek to show how well industry is complying with environmental expectations. The following points are the general talking points written consulting forms for industry dealing with precautionary principles. (www.mindfully.org/Precaution/Precautionary-Principle-wirthlin.htm)

1. It is too late to redefine the precautionary principle in ways favorable to industry. Instead, emphasize industry's adherence to a tried and proven precautionary approach, and draw the distinction between reasonable and extreme interpretations.
2. Enlist surrogates who can effectively attack the precautionary principle and its misapplications. Language and positioning are all-important. Beware of saying that industry opposes the precautionary principle. Activists will seize on that to imply that big business does not care about people's health and safety.
3. Drive up the negatives associated with the precautionary principle.
4. Find examples within your own industry to demonstrate the obvious dangers of relying on an over-simplified version of the principle.
5. Build support for traditional risk assessment

Industries should conduct research and message testing to determine how to rebrand risk assessment in order to make it salient to consumers and compelling enough to blunt the rhetoric of Precautionary Principle advocates. This requires proactively orchestrating support for the science-based process of risk assessment by influential people in the public and private sectors. The following are possible message themes:

- A science-based precautionary approach, known as risk assessment, has been a part of U.S. environmental and health laws for over 100 years. We don't need the Precautionary Principle.
- Risk assessment has served the United States well. American life expectancy nearly doubled during the 20th Century. Americans today are living significantly longer and healthier lives than ever before. Industry is committed to a science-based precautionary approach that safeguards human health and the environment. This approach has been very successful in protecting consumers from unsafe products while respecting the consumer's freedom of choice in the marketplace.
- Industry invests heavily in safety testing and product improvement. Companies who make such an investment have a huge economic incentive not to put on the market a product that will later be shown to be unsafe.

6. Conduct core-values research among consumers and other key stakeholders, to understand their perceptions (both rational and emotional) of your industry on health and environmental issues, and to anticipate their likely response to public debate over the Precautionary Principle.

7. Develop a communications strategy and employ message testing to identify how to most persuasively communicate your position on these issues as appropriate within your company, to regulatory bodies, to industry peers, to customers, and to the general public.

Because of the rapid growth in the application of the precautionary principle and its potential cost to industry, conflict and controversy are inevitable.

POTENTIAL FOR FUTURE CONTROVERSY

A precautionary approach has many feedback loops with new and engaged stakeholders, like communities. As uncertainty is reduced about the harmfulness of a given activity better environmental decisions can be made. This implies that all stakeholders should have access to relevant information in a public participation process far more extensive than that currently practiced in the United States. Activities that pose too many uncertainties and offer too few benefits will increasingly be rejected. Slow regulatory processes and the financial risk of a no decision, combined with expensive scientists and exposure to litigation, are costs borne by industry. Advocates of sustainability strongly support the application of the precautionary principle to all decisions that affect the environment. Right now the momentum for application of the precautionary principle is moving fast globally and locally, but it is still very new. It is likely that controversies will emerge in city councils, county commissions, and state legislatures as industrial and business interests respond to this new, environmental decision-making process.

See also Climate Change; Cumulative Emissions, Impacts, and Risks; Ecosystem Risk Assessment; Global Warming; Human Health Risk Assessment; Public Involvement and Participation in Environmental Decisions

Web Resources

Rachel's Democracy and Health News. 2007. Toxicants in Synthetic Turf. Available at www.rachel.org/bulletin/index.cfm?issue_ID=2426. Accessed January 21, 2008.

Rachel's Environment and Health News. 1998. Wingspread Statement on the Precautionary Principle. Available at www.mindfully.org/Precaution/Precautionary-Principle-Rachels.htm. Accessed January 21, 2008.

Tickner, Joel, and Nancy Myers. 2000. Precautionary Principle: Current Status and Implementation. *Synthesis/Regeneration* 23. Available at www.greens.org/s-r/23/23–17.html. Accessed January 21, 2008.

Further Reading: Common, Michael S. 1995. *Sustainability and Policy: Limits to Economics.* Cambridge: Cambridge University Press; O'Riordan, Timothy. 2006. *Interpreting the Precautionary Principle.* London: James and James/Earthscan; Sunstein, Cass R. 2005. *Laws of Fear: Beyond the Precautionary Principle.* Cambridge: Cambridge University Press; Tickner, Joel A. 2002. *Precaution, Environmental Science, and Preventive Public Policy.* Washington, DC: Island Press.

PRESERVATION: PREDATOR MANAGEMENT IN OREGON

Wildlife advocates claim that the problem of political influence overruling species management can be seen in the historic persecution of mountain lions in the state of Oregon, where the Oregon Department of Fish and Wildlife (ODFW) and conservationists face each other in a classic controversy.

Western states with large federal landholdings are often battlegrounds for environmental preservation. Ranchers, farmers, indigenous people, and environmentalists all fight for their interests here. State wildlife agencies are the modern-day battlegrounds for these controversies.

Extensive predator-eradication campaigns throughout the twentieth century were conducted for the convenience of ranchers and the funds that could be engendered by the sale of kill tags. The last grizzly bear in Oregon was killed in 1934. Now the mountain lion, or cougar, is the only remaining predator species in the state. Conservationists point to the state's mismanagement that almost eradicated every cougar in the state of Oregon by the middle of the last century. State bounties of $5 were paid to houndhunters for a cougar's tail. The slaughter was so egregious that by the 1970s, when the bounties ended, biologists estimated that perhaps as few as a dozen cougars were left in the whole state by time the state's predator-eradication campaign ended. Yet by the 1980s ODFW wanted to start allowing houndhunters to kill them again.

Beginning in the late 1980s, wildlife activists and conservationists all over the country spoke out about what they claimed was mismanagement of large predators for the benefit of a few vested interests. Conservation groups like Predator Defense and the Northwest Cougar Action Trust repeatedly testified before Oregon's Wildlife Commission and the legislature, protesting what they perceived as regressive predator-management policies. Wildlife advocates protested that the state had no biological basis for opening up renewed hunting seasons on predators. Political pressure from groups like the National Rifle Association (NRA) and other hunting proponents led the ODFW to raise the kill quotas on cougars each year through the 1980s and 1990s.

In Oregon, a successful citizens' initiative was passed in 1994. Measure 18 banned the then-popular hunting techniques of baiting and hounding bears and cougars, with packs of radio-collared dogs. Then state game officials protested that voters had taken away some of their management options and worked to encourage hunting groups to try and repeal the voter-mandated limits on the killing of cougars and bears. It was at this pivotal point where the International Safari Club started instigating a push for renewed permission to hunt Oregon's lions for trophies. These groups openly funded several failed attempts to reverse voter-mandated protection for mountain lions.

In 1996 they tried to use the initiative process to reverse the law; but they failed again. State game officials still do not seem to understand that the reason their efforts at predator eradication continually fail, is that apparently the majority of voters still believe that cougars and bears should not be hounded or baited in what the public considers a cruel, unfair fashion. Conservationists say that the state's one-sided management philosophy has shaken the public trust in wildlife managers. But to this day it is still illegal to harass wildlife with packs of tracking dogs in Oregon, unless there is a direct threat to human life or property.

Game officials defend the renewed use of hounds as the only viable method of limiting lion populations. They contend that reports of cougar sightings are increasing and say renewed hounding is necessary for the protection of the public and ranchers. Conservations say that cougars pose a small public safety threat. They point to the fact that the chance of being attacked by a cougar is less than that of being hit by lightning (1 in 1,000,000). Biologists know that humans are not on the menu. In fact, mountain lions are such effective hunters that if they really did want to attack us, they would easily have the upper hand. Conservationists refute the state's argument for reintroducing these banned practices as not defensible since the public safety threat is small, and public education campaigns can be provided to improve husbandry practices. Researchers point out that the overwhelming preponderance of such agricultural losses are caused by coyotes, not cougars. Advocates say strong public sentiment runs in favor of state reimbursement programs for ranchers' losses, and leaving the lions alone.

Wildlife advocates say it is discouraging that even after decades of public protests about outdated predator policies, the state remains recalcitrant and persists in demanding that they be allowed to go directly against the voters wishes and reinstitute the same practices that almost killed off every lion in Oregon.

Conservationists maintain that several successful legal voter initiatives corroborate the strong public sentiment for protecting lions, and they accuse state game officials of catering to a small, economically insignificant segment of Oregon's population. They point out that fee-paying hunters make up less than 5 percent of Oregon's population. Other, more progressive states advocate public education outreach to teach agricultural interests ways they can improve their animal husbandry practices.

In 2006 the Oregon Department of Fish and Wildlife approved a new cougar plan and, as part of it, requested that the U.S. Department of Agriculture's Wildlife Services program start implementing a new predator eradication plan, to placate ranching interests. But the federal Wildlife Services decided to start killing cougars without first preparing an assessment of the environmental effect of those killings, a violation of the National Environmental Policy Act. The state's new cougar plan calls for the killing of between 2,000 and 3,000 cougars in broad areas of the state. The ODFW's stated goal is to "manage for a cougar population that is at or above the 1994 levels." The plan was immediately lambasted by conservation groups that claim it was more of a cougar-killing plan than biologically defensible policy. One outspoken group wrote the governor that ODFW is knowingly using faulty data to promote its newly revised cougar plan. The ODFW has solicited reports of cougar incidents and then used them to fuel public hysteria with claims of dramatic increases in complaints and sightings.

The department has continued to assert that Oregon's cougar population is exploding despite the research of expert scientists who refute the figures. The ODFW has used the media to promote a fear campaign.

In June 2006, the Humane Society of the United States and six other conservation organizations filed suit in the U.S. District Court for Oregon to stop Wildlife Services from killing cougars, under the state's direction. They maintain that the whole Cougar Management Plan is flawed, because it is predicated on incorrect and obsolete data. Conservationists claim that under the guise of "management," this plan will allow federal employees to re-introduce houndhunting and indiscriminately kill cougars in broad areas of the state, resulting in biologically unsound local extirpations. The state still maintains that it intends to try and freeze predator populations at some arbitrary point in the past history.

Wildlife advocates maintain that the ODFW has fabricated its statistics and that their plan has no biologically defensible basis. They point out that the proposed plan is obviously constructed to placate ranchers and is not supported not by empirical data. Additionally, advocates testified that besides the fact that the state has no conclusive published evidence that cougars might be the cause of elk declines, only published data on cougars show an astounding mortality rate for the predators. Studies both east and west of the Cascades prove that lion populations are still directly threatened by overhunting and habitat degradation. When the state finally did put collars on cougars, what they documented was actually rather stark proof of how heavily cougars are being hunted and poached in Oregon.

There is no direct connection between the number of complaints and the actual cougar population. Oregon's human population has doubled in the last 20 years. So, humans have invaded the lions' space in many areas; mere sightings do not mean that there are more cougars. When there are more people in a given area, authorities should expect a corresponding increase in reported sightings.

They have constructed hunting incentives for the killing of cougars all year long. They continue to sell unlimited kill tags for cougars that can be purchased by hunters for $10 as part of their harvest quota system, the chief economic support of the ODFW. There is effectively no law enforcement in remote areas. The state knows that this wide-open permission to kill lions on sight will directly result in their widespread killing, virtually all year long, while relying on a system of voluntary compliance.

So, predator advocates still continue to fight a running battle for protection of this invaluable species. The Humane Society and other conservationists continue to strongly oppose the state's mismanagement and new cougar plan, commenting that it is full of arbitrary regulations and unrealistic assumptions. They have characterized it as a travesty of a wildlife plan. They also want to alert the public to what they feel is an ongoing threat to lion populations in the West, wealthy game hunters with professional guides who help the trophy hunters win points for killing certain animal species that count toward their Boone & Crockett Grand Slam score.

States like California, which have not allowed any hunting of lions for decades, realize that rural residents need to be educated and that it is really a question of social tolerance levels. Heavy resource extraction on federal lands throughout the West has created an extensive system of roads, crisscrossing much of the land base. There are currently no real large blocks of roadless habitat that are protected from vehicular and off-road motorists in Oregon. Cougar and elk biologists say these animals urgently need adequately protected migratory habitats with high-quality forage.

POTENTIAL FOR FUTURE CONTROVERSY

As human populations threaten species habitat, more controversies about species preservation will arise.

See also Endangered Species; State Environmental Land Use

Web Resources

California Department of Fish and Game. Keep Me Wild. Available at www.dfg.ca.gov/keepmewild/lion.html. Accessed January 21, 2008.

Humane Society of the United States. Available at www.hsus.org. Accessed January 21, 2008.

The Mountain Lion Foundation. Available at www.mountainlion.org/index1.asp. Accessed January 21, 2008.

Oregon Department of Fish and Wildlife. Key Facts about Cougars in Oregon and the Cougar Management Plan. Available at www.dfw.state.or.us/wildlife/cougar/. Accessed January 21, 2008.

Safari Club International. Record Book Top Ten and Minimums. Available at www.safari-club.org/members/RB/view_minimums.cfm. Accessed January 21, 2008.

Washington Department of Fish and Wildlife. Game Management Plan, draft. Available at wdfw.wa.gov/hab/sepa/draft_gmp_18oct02.pdf. Accessed January 21, 2008.

Further Reading: Botkin, Daniel B. 1992. *Discordant Harmonies: A New Ecology for the Twenty-First Century.* New York: Oxford University Press; Moehring, Eugene P. 2004. *Urbanism and Empire in the Far West, 1840–1890.* Reno: University of Nevada Press; Porter, William F., Joseph L. Sax, and Frederic H. Wagner. 1995. *Wildlife Policies in the U.S. National Parks.* Washington, DC: Island Press; Pritchard, James A. 1999. *Preserving Natural Conditions: Science and the Perception of Nature.* Lincoln: University of Nebraska Press; Rutherford, Murray B., and Denise Casey, eds. 2005. *Coexisting with Large Carnivores: Lessons from Greater Yellowstone.* Washington, DC: Island Press.

Cathy Koehn

PUBLIC INVOLVEMENT AND PARTICIPATION IN ENVIRONMENTAL DECISIONS

Public participation in some environmental decisions is mandated by law. This participation is very limited and criticized as noneffectual. It is basically minimal notice of the right to see documents or place something in the record. The public expects to have notice and involvement in every important environmental decision that affects where they live, work, and play.

THE AARHUS CONVENTION

International public involvement and participation in environmental decisions and international perspectives on the environment tend to be more holistic and inclusive than U.S. concepts. Water quality in urban areas has long been an international environmental priority but is only recently emerging as one in the United States, for example. What are the international inspirational goals and ideals about involving the public with environmental decisions? What kind of process do they want to use based on their 500- to 2,000-year history of European civilization?

> Although regional in scope, the significance of the Aarhus Convention is global. It is by far the most impressive elaboration of principle 10 of the Rio Declaration, which stresses the need for citizens' participation in environmental issues and for access to information in the area of environmental democracy so far undertaken under the auspices of the United Nations. (Kofi A. Annan, Secretary-General of the United Nations)

The UNECE Convention on Access to Information, Public Participation in Decision-making and Access to Justice in Environmental Matters was adopted on June 25, 1998, in the Danish city of Aarhus at the Fourth Ministerial Conference in the Environment for Europe process. The Aarhus Convention is a new kind of environmental agreement.

- It links environmental rights and human rights;
- It acknowledges that we owe an obligation to future generations;
- It establishes that sustainable development can be achieved only through the involvement of all stakeholders;
- It links government accountability and environmental protection;
- It focuses on interactions between the public and public authorities in a democratic context; and
- It is forging a new process for public participation in the negotiation and implementation of international agreements.

The subject of the Aarhus Convention goes to the heart of the relationship between people and governments. The convention is not only an environmental agreement; it is also a convention about government accountability, transparency, and responsiveness.

The Aarhus Convention grants the public rights and imposes on parties and public authorities' obligations regarding access to information and public participation and access to justice.

Environmental public-participation policies create controversy because they deal with a public that often distrusts the process. The issues can be technically complex, perceived as life threatening, and value-laden. A number of research findings underscore the importance of improving public involvement in environmental decision making. The emerging policy consensus is that public involvement can help develop better knowledge of the ecology and culture of the place. The lack of land-use planning of an environmental nature at the local level in the United States leaves a blank gap in the knowledge of past and present land uses and their environmental impacts. What land-use planning for the environment that does exist seldom goes beyond big parks and riparian greenbelts. Most local environmental planning is just beginning to get beyond protection against natural disasters, like floods. Even at its current best, most local land-use planning has poor to nonexistent relationships with federal and state environmental agencies. Environmentalists accuse them of a promoting development over the environment by letting industry use the lack of information at the intergovernmental level. Recent changes in citizen access to the sets of memorandums of understanding (MOUs) that exist between the federal regional offices of environmental agencies and the states, environmental agencies help give citizens knowledge about revenue flow and policy requirements for different environmental programs, many related to environmental enforcement and compliance. This helps with the state and federal governmental relationship and its transparency but does nothing for local environmental planning and its relationship with the other two government levels and their agencies. This has been and will continue to be a large battleground exposed by increased citizen involvement. It is a rapidly changing dynamic, and the battle goes in different directions in some state legislatures. Agencies have unresolved policy issues of efficiency to contend with in their own performance. How much public participation is too much or too little? Overall, more citizens are seeking involvement in environmental decisions whenever they can. Citizens are generally not accommodated, and there are many anecdotes about restricted public meetings. States are reluctant to provide childcare at public hearings in some states for fear of liability. This prevents many working parents with environmental concerns from attending meetings. In Hawaii environmental public hearings can be held on different islands. Travel between the islands generally requires air travel, which can be too expensive for many with environmental concerns. Hearings can be held at night, during the day, and without notice. Public hearing format and agenda form a battleground that is site specific. The hearings can be recorded, handicapped accessible, with language and deaf translation, and provide capacity-building trainings. Many communities are at first overwhelmed with the scientific complexity of some environmental issues. Some people come to the hearings to get information or to give information. The struggle for information can often set the early parameters of a given battleground. This is especially the case when the information conflicts with neighborhood realities.

The public-participation requirements of environmental law to involve the public, such as informal and formal comments, public hearings, public comments, and citizen suits, have proved inadequate to effectively meet the challenge of constructively involving the public. Many agencies now develop citizen advisory groups. The Environmental Protection Agency, Department of Energy, and Department of Defense have more than 200 citizen advisory groups at contaminated sites around the country; advisory groups have become important components of EPA's environmental justice activities. Some manufacturing industries have plants that use citizen advisory committees. Yet public involvement is often part of any controversial environmental issue. The real issue is often an environmental conflict, such as the failure of enforcement of pollution laws, which affects the community. Some of the battlegrounds are defined by the goals of the public participation. Is a citizen involvement program successful if it simply involves more of the community, or should it have to result in demonstrably better decisions in terms of overall environmental impacts? This question tugs at many of the internal agency battles to go beyond legal requirements for citizen participation.

Another battleground of this controversy arises from fundamental differences about citizens' role in government. Most citizens assume that they have a right to participate in the large-scale environmental decisions that affect them and their private property. There is a large difference in how that involvement and participation actually happens. Are there many meetings? Are they offered in other languages? How much power over the decision should citizens have? Often citizens think they have the power to stop an environmentally hazardous use, but they do not. Citizens' opinions are noted, but the administrative agency makes the decisions. After an unsatisfactory decision citizens can appeal through the agency, and after that they can go to court. Industry is often held captive by the agency as the agency deals with citizen participation and involvement. The direct involvement of citizens in decision making is seen as a threat because it opens the door to self-interested strategic behavior. The participation requirement in environmental decision making is more than in most other areas of government. It is required by many environmental laws. These processes can be value-laden and explosive when cultures clash. For example, many administrative agencies diligently prepare meeting agendas in advance of the meeting. Sometimes they involve the community. However, when citizens attend the hearing, they may deviate from the agenda, considering it just one stakeholder's view of the appropriate topic of discussion. Citizen involvement processes are hard for many administrative agencies because there are no criteria for judging success and failure. This too contributes to the controversy around public participation and involvement.

DEVELOPING ISSUES WITH PUBLIC INVOLVEMENT

Many of the problems of public involvement revolve around the capacity of the public to engage in the decisions. It may take longer to reach a decision or formulate a process because of the need for community capacity building, group processes, and the expectation that government and industry be able to meaningfully answer all questions and concerns. The U.S. public lack basic knowledge about many environmental issues but quickly become engaged when there is environmental controversy.

OBJECTIVES OF PUBLIC PARTICIPATION

In U.S. environmental policy and law, public participation has some basic goals. The goals are:

- Educating and informing the public about a prospective environmental decision
- Improving the substantive quality of decisions
- Giving notice and a meaningful opportunity for all interested parties to participate
- Reducing controversy, especially in environmental impact assessment cases

WHAT ARE PROCESSES OF PUBLIC PARTICIPATION?

A generalized lack of democracy in environmental decision making is the core of the problem. Environmental decision making brings citizens into the complex world of regulation, industry, and science.

The basic mechanisms of citizen participation in the United States include:

- Public hearings, public meetings, roundtable meetings, public comments, and advisory committees
- One-way flows of information such as surveys, focus groups, and public education
- Citizen involvement techniques with collaborative decision making and conflict resolution, such as mediation

ROLE OF ENVIRONMENTAL LITERACY

Environmental literacy allows citizens to carry out the role envisioned in major environmental laws such as the Clean Air Act and Clean Water Act. Laws such as the Toxic Release Inventory go a long way toward increasing environmental literacy because they act on self-interest. Citizens can find out the emissions of some polluters in their zip code or neighborhood. Environmental literacy helps citizens' ability to participate in decision making. Ideally, the citizens' increase in their environmental capacity will enable them to deliberate issues with government agencies, industries, and scientists.

QUESTIONS TO ASK ABOUT PUBLIC PARTICIPATION

Many times community members and environmentalists learn about a certain meeting or decision after the public-participation phase took place. The most common reason for this is the inadequacy of the notice to people or the failure of the government to designate that member as an interested party. Here are some common questions asked to determine the nature and validity of a given public-participation process.

- How many members of the public were actively involved in participatory venues or took advantage of information and access provided to them?
- What percentage of the wider public was reached through education campaigns, media relations, or interaction with more active participants?

- Did the active public feel that they had sufficient knowledge to contribute to deliberations and decision making?
- Did members of the public understand their role in the participatory process?
- Was there sufficient time and money available to obtain credible, relevant, and, if necessary, independent information?
- Was information from the public-participation process used to inform or review analyses or decisions?
- Did the public feel that it had an impact on decisions?
- Where public input was not incorporated into analyses or decisions, did the relevant agency provide justification that was acceptable to the public?
- Were all reasonably affected parties included or represented, particularly those with no formal organization?
- Did participants reflect the larger public they were expected to represent, for example, in terms of socioeconomic criteria?
- Were there mechanisms to hold participants accountable to the community that they represented?

Citizens are an important repository of observations, facts, and innovative alternatives.

The goal of cost-effectiveness does not refer to the cost-effectiveness of decisions made in participatory processes but to the cost-effectiveness of choosing among the different participatory or nonparticipatory approaches to environmental decision making. The agency perspective argues that public participation programs must be cost-effective by producing results that justify the added effort of including citizens. They face the very pragmatic issue of how much citizen participation is too much or too little. Others argue against this view because it is too difficult to measure the value of environmental harm that is prevented. Others say that the timescale used measure the effectiveness of the environmental intervention is more determinative. Over a longer timescale citizen participation can monitor and adapt the environmental decision to changing conditions, for instance. On a short timescale the citizen participation may be more costly and the benefits less measurable.

Public hearings remain the most common form of face-to-face public involvement. The Environmental Protection Agency convenes thousands of hearings per year. Most are used to defend EPA decisions rather than to involve the public in the decision-making process itself. They are judicial or quasijudicial in nature and generally precede a court case.

Environmentalists and communities complain that state and federal agencies often hold hearings too late in the process, intentionally present technical information beyond the understanding of the lay public in a language that is hard to understand, and minimally fulfill legal requirements. They complain that the hearings are like theater. The decision is already made, they contend. The process is not deliberative, and their contributions and concerns are not meaningfully acknowledged. Once the decision is made and is final, the citizens may be able to legally contest it.

POTENTIAL FOR FUTURE CONTROVERSY

The further the public is kept away from environmental decisions that affect where they live, work, or play, the more controversial the decision will become. As U.S. mechanisms for public participation are rudimentary, more sophisticated models of public involvement will probably follow international models.

See also Citizen Monitoring of Environmental Decisions; Environmental Impact Statements: United States; Good Neighbor Agreements; Sustainability

Web Resources

U.S. Environmental Protection Agency: Public Involvement. Available at www.epa.gov/publicinvolvement/. Accessed January 21, 2008.

U.S. Environmental Protection Agency. Wastes: Public Participation and Citizen Action. Available at www.epa.gov/epaoswer/hazwaste/permit/pubpart/. Accessed January 21, 2008.

Further Reading: Depoe, Stephen P., and John W. Delicath. 2004. *Communication and Public Participation in Environmental Decision Making.* Albany, NY: SUNY Press; Krimsky, Sheldon, and Dominic Golding, eds. 1992. *Social Theories of Risk.* Westport, CT: Praeger; National Environmental Justice Advisory Council (NEJAC). 1996. *The Model Plan for Public Participation.* Washington, DC: Environmental Protection Agency; U.S. Environmental Protection Agency (EPA). 1995. *Guidance for Community Advisory Groups at Superfund Sites.* EPA 540-K-96–001. Washington, DC: Office of Emergency and Remedial Response, U.S. EPA.

R

RAIN FORESTS

Rain forests cover about 6 percent of the earth's surface. Rain forests are evergreen woodland forests with heavy precipitation, about 100 inches per year, and a continuous canopy of leaves. The range and depth of their biodiversity is enormous. They are home to over half the species of plants and animals in the world. They convert large amounts of carbon dioxide into oxygen. Deforestation of the rain forest is an international environmental controversy. Deforestation comes from logging, mining, slash-and-burn agriculture, and ranching developments.

Rain forests were often considered the raw jungle. The popular perception was one of dense and tangled tropical growth interspersed with many unknown and dangerous insects and animals. Most forests were regarded either as wood for fuel and building or as impediments to human habitation. Industrial society and European colonization used forests as free sources of valuable materials or as woods, occupying land and getting in the way of development. Most rain forests are now cleared by a much more powerful technology for their timber. Tropical hardwoods can command a high price on the world market. In the rain forests logging is usually followed by farming and ranching operations. Often they are not local operations but an extension of multinational corporations.

In the late 1970s deforestation was widely recognized as an environmental issue of global importance. In 1992, with much controversy between developed and developing countries, the United Nations reached the first global consensus on forests and deforestation. The outcome was the "Non-Legally Binding Authoritative Statement of Principles for a Global Consensus on the Management, Conservation and Sustainable Development of All Types of Forests" (the "Forest Principles"). This was also part of Agenda 21, which was the program for sustainable development from the first Rio Earth Summit, dealing with combating deforestation. Global resistance to deforestation has since become more entrenched, partially because of reports of climate change due to global warming. Both the Intergovernmental Panel on Forests (IPF) and the Intergovernmental Forum on Forests (IFF) have been established under the auspices of the United Nations Commission on Sustainable Development.

BATTLEGROUND: DOCUMENTING BIODIVERSITY

Rain forests exude life and biodiversity in most of their ecosystems. More than 50 percent of the earth's known species live in tropical rain forests. A representative four-square-mile patch of tropical rain forest contains up to 1,500 species of flowering plants, 750 tree species, 125 mammal species, 400 bird species, 100 reptile species, 60 amphibian species, and 150 butterfly species. At least 25 percent of all modern drugs originally came from rain forests. Over 2,000 tropical plants have been identified by scientists as having anticancer properties. Many parts of the rain forest remain unexplored by scientists. Historically rain forests covered 14 percent of the earth's land surface. It is now estimated they cover 6 percent of the planet's land surface. Some experts estimate that the last remaining rain forests could be consumed in less than 40 years. Experts estimate that 137 plant, animal, and insect species are destroyed every day due to rain forest deforestation. Many fear that species of plants and animals will go extinct before they are scientifically categorized. One battleground in this controversy is simply documenting the rich biodiversity of the rain forest. Work on this task is proceeding at an accelerating rate, aided by technology. It is still far from complete. One obstacle to this task is the next battleground in this controversy.

BATTLEGROUND: ENVIRONMENTAL EXPLOITATION?

Some accuse pharmaceutical research companies of mining the rain forest for medical plants. The companies then get the patent or copyright to it. They negotiate with both countries and indigenous people. Currently, 121 prescription drugs sold worldwide derive from plant sources. About 25 percent of Western pharmaceuticals are from rain forest ingredients, and less than 1 percent of these tropical trees and plants have been tested by scientists. In 1983, no U.S. pharmaceutical manufacturers were involved in research programs to discover new drugs or cures from plants in the rain forest. Today, over 100 pharmaceutical companies, several universities, and the U.S. government are engaged in plant research projects for possible drugs and cures for many modern-day afflictions. There is a fear in the scientific research community, both public and private, that valuable medical information could be lost due to rain forest deforestation.

OVERVIEW OF THE RAIN FORESTS

There are rain forests all over the world, in many different countries. The battlegrounds over rain forests are often specific to the particular country.

Central America was once completely rain forest, but large areas have been cleared for cattle ranching and for sugar cane plantations. Similar to other major rain forests, the jungles and mangrove swamps of Central America contain many plants and animals found nowhere else. Central America has large numbers of tropical birds, including many kinds of parrots. The Amazon watershed is the world's largest tropical rain forest. The forest covers the basin of the Amazon, the world's second-longest river. The Amazon contains the most biodiversity on earth, as far as it is known to the present. One-fifth of the entire world's plants and birds and about one-tenth of all mammal species are found there.

Africa is a large continent with large areas of rain forests and unknown biodiversity. Africa also contains areas of high cloud forest, mangrove swamps, and flooded forests. Central Africa holds the world's second-largest rainforest. To the southeast, the large island of Madagascar was once intensively forested, but now much of it is gone. The island of Madagascar is home to many unique plants and animals not found anywhere else.

The rain forests of Asia stretch from India and Burma in the west to Malaysia and the islands of Java and Borneo in the east. Bangladesh has the largest area of mangrove forests in the world. Mangrove forests play an important role in many coastal ecosystems.

Australia, New Zealand, and the island of New Guinea contain many different species of animal that occur nowhere else. Undergrowth in Australia's tropical forests is dense and lush. The forests lie in the path of wet winds from the Pacific.

RAIN FORESTS AND INDIGENOUS PEOPLES

The population of indigenous people in the Amazonian rain forest decreased from an estimated 10 million to fewer than 200,000 today. European colonists destroyed many indigenous tribes since the 1900s. Some of these tribes and bands hold treaty rights to the land but have difficulty getting the provisions of the treaty enforced locally. These groups could decide to economically develop the land by removing the forests. One concern here is the loss of traditional ways of medicine. The medicine man of these groups often possesses centuries of knowledge of indigenous plants. Many of these medicine men are now more than 70 years old and are not being replaced. There are several small projects under way to try to capture some of the oral history generally, some by the tribes themselves.

The vast bulk of the world's rain forests are in developing countries, many of which also have weak control by government. Timber and land are valuable resources that cannot be ignored. Much of the pressure for conservation comes from the developed world. In the developed world the economic issue is less directly significant than the environmental and cultural issues. The cultural clash of values between developed and undeveloped nations, between indigenous peoples and scientific explorers, and between global environmentalists and multinational corporations outlines the battleground of the rain forest deforestation controversy.

THE WEALTH OF THE RAIN FORESTS

The Amazon rain forest covers over a billion acres of land. It includes parts of Brazil, Venezuela, Colombia, and the eastern Andean region of Ecuador and Peru. More than 20 percent of the planet's oxygen is produced in the Amazon rain forest alone. More than half of the world's estimated 10 million species of plants, animals, and insects live in the tropical rain forests. One-fifth of the world's fresh water is in the Amazon Basin, one of the largest and oldest intact watersheds on the planet. It is here that slash-and-burn land settlement techniques destroy many acres of rain forests.

POTENTIAL FOR FUTURE CONTROVERSY

The scientific evidence for the role played by the rain forests in supporting life on earth is no longer disputed. Deforestation destroys biodiversity and indigenous cultures, leads to desertification and land degradation, and contributes to the global warming by reducing the planet's capacity to recycle carbon dioxide. These are all former battlegrounds.

While international agreements are made to try to preserve rain forests as well as respect state sovereignty and cultural traditions, there are still issues about how well they are enforced. Poor countries do not want to give up a natural resource like wood just because rich countries want to preserve rain forests. Some countries, such as Costa Rica, have capitalized on the beauty of the rain forest through successful ecotourism. Other countries may not have this option. Deforestation is an international problem, but deforestation of the rain forest affects the planet most. The emerging battleground for this controversy may be the area of international environmental law enforcement.

See also Climate Change; Conservation in the World; Ecotourism as a Basis for Protection of Biodiversity; Logging

Web Resources

Rainforest Facts. Available at www.rain-tree.com/facts.htm. Accessed January 21, 2008.
World Rainforest Information Portal. Available at www.rainforestweb.org/. Accessed January 21, 2008.

Further Reading: McClain, Victoria, Richey, eds. 2001. *The Biogeochemistry of the Amazon Basin.* New York: Oxford University Press; Rajala, Ruchard A. 1999. *Clearcutting the Pacific Rain Forest.* Vancouver: UBC Press; Rudel, Thomas K. 2005. *Tropical Deforestation: Small Farmer and Land Clearing in the Ecuadorian Amazon.* New York: Columbia University Press.

S

SACRED SITES

Sacred sites are places revered by local and indigenous people for their spiritual histories and qualities. They include burial grounds and geological locations thought to be powerful or sensitive in terms of metaphysical forces. These sites are found in all cultures, in all countries, and among all religions. Sacred sites are often involved in environmental controversies over resource development or tourism, as opposed to preservation of indigenous or ancient cultures and biodiversity. Intense pressure to develop timber, oil and gas reserves, coal mining, and nuclear waste storage have all intensified developmental pressures on sacred lands. These sites can stop development because any significant impact on them requires an environmental impact statement.

CONTROVERSIES WITH NATURAL RESOURCE EXTRACTION

Many native lands, once thought to be without exploitable resources, are under intense scrutiny from gold, coal, and uranium mining companies. For example, the ancestral homelands of eight Lakota Sioux nations, the Black Hills of South Dakota, are now valued at more than $4 billion and are the subject of an unpaid claim settlement against the U.S. government for $570 million. These lands are considered the womb of the people of these tribes and cover lands in contemporary western South Dakota, northeastern Wyoming, and southeastern Montana. Without participation from affected tribes, logging interests, sports and recreational interests, and others have all been allowed great access and use of these lands despite resistance from organizations such as the American Indian Movement.

These tribes continue to advocate the return, undisturbed, of their lands, refusing settlement monies despite some of the lowest incomes and employment rates in the country. Cyanide heaps and leached gold mines have left lands and water polluted, and ecotourist developers have proposed extensive development on these lands despite objections regarding these users' disrespect of spiritual customs on the land.

In the western states, the competition for freshwater resources to feed growing populations and agribusiness has led to a proposed public works project to increase the height of the Shasta Dam on the McCloud River, flooding the remaining ancestral homeland and sacred sites of the Winnemem Wintu. Similar struggles pitting indigenous people against coal-mining interests and dam-building projects have arisen in the two fastest-developing economies in the world, India and China. In India, coal mining threatens the valley of the Damodar containing invaluable archeological sites and sites sacred to several indigenous tribes. In China, the Three Gorges Dam on the Yangtze River in Sichuan Province threatens both sacred sites and biodiversity in the region, which are being sacrificed to the need for a source of energy to fuel the developing economy.

TOURISM

Some sectors of the tourism industry have had a long history of conflict with indigenous tribes and their claims of sovereignty and sacredness of land that presented recreational opportunities like ski resorts and rafting adventures. Mount

---

HIKING ON AYERS ROCK, AUSTRALIA

In the middle of the Australian continent, a series of vivid red rock domes rise some 1,100 feet from the desert floor, making a visually striking image that has attracted visitors and indigenous people alike. The largest of these rocks are called Uluru and Kata Tjuta by the indigenous Anangu people; they are known to western tourists as Ayers Rock and Mount Olga. The government of Australia developed them as tourist destinations, but they have now been restored to the stewardship and joint ownership of the Anangu people by the Northern Territory Aboriginal Sacred Sites Act of 1989. Uluru and Kata Tjuta are sacred to the Anangu people: they hold more than 40 named sacred sites and countless other secret sites, as well as numerous *iwara* (ancestral paths) crisscrossing the area. The Anangu did not climb on Uluru (Ayers Rock), but it remains a very popular tourist attraction for hikers, backpackers, and climbers despite requests not to climb or hike there. There are numerous Web sites offering this experience without any reference to the wishes of the indigenous stewards of this land. Moreover, climbing the rock is dangerous. At least 37 people have died while making the climb since tourism has operated there. For each death, the Anangu are obliged to grieve. For these reasons, Anangu request that visitors do not climb the rock. But about 400,000 people visit the park annually, and many choose to attempt the hike, creating multiple strains on cultural traditions and infrastructure.

Shasta in northern California and the San Francisco Peaks in Arizona are considered holy ground by their local tribes but are also the sites of proposed multi-million dollar ski resorts. Increasingly, tourists seeking spiritual experiences and ecologically fulfilling travel destinations have also sought out esoteric and exotic destinations including sacred sites. Ecological tourism, called ecotourism, has emerged as a theoretical compromise promoting travel-related services and development aimed at both ecologically and socially conscious tourists. The concept is based on the idea that tourist dollars can be channeled in ways that benefit the environment and local people without creating the environmentally degrading consequences normally associated with recreational development, while allowing more equitable local participation in development benefits.

LOCAL AND INDIGENOUS COMMUNITIES' CLAIMS

Underlying many of these controversies is the fact that indigenous societies have traditionally regarded land and its use fundamentally differently from a western idea of private ownership and exclusive rights to exploit land. Communally held lands and sacred lands were not to be held or exploited for the benefit of individual wealth creation, and treaties executed between colonial or invading governments and indigenous peoples were often based on fundamental misperceptions of basic terms. What European settlers meant when they negotiated the sale of land had no comparable meaning in the minds of those tribes with whom they dealt. Many tribes simply had no concept that individuals could own land; they conceived of the relationship between human and land as the other way around—humans are the people of the earth. Even more to the point, some land could not be used at all, for any purpose except ceremonial and secret purposes, intended to benefit the tribal society and tribal people. These indigenous limits on the use and ownership of land had no parallel in the European settler's law.

Contemporary treaty interpretations have repeatedly stumbled over ways to resolve the inequities of these fundamentally mistaken documents. There are controversial issue of sacredness and proof of claims, secrecy issues, and western concepts of evidence. Tribes often claim rights in these areas and simultaneously claim the solemn obligation not to reveal the basis for their claims, refusing to submit their claims to outsiders' review. Often there is no paper trail of ownership such as that traditionally used by western European settlers. Nonetheless, in some island nations land can be inherited to the 10th cousin. It is so precious it is not bought or sold. Outsiders, often the majority culture, are suspicious of these claims and processes of proof. These controversies flare up when the land in question becomes valuable.

POTENTIAL FOR FUTURE CONTROVERSIES

As human population increases and development moves more into indigenous lands all over the world, sacred sites may become more imperiled. Whether an object of tourism or a deterrent to development, sacred sites represent a

requires large amounts of water. The water comes from the surrounding streams and creeks on the mountain and other sources. Snowmaking requires withdrawal of water from these sources. The biggest pressure for water for snowmaking equipment is to open the skiing season as early as possible, in the late autumn or early winter in most places. Summers can be very dry, so some of these water sources can be very low when the demand for the water is high for snowmaking. This can create a battleground between downstream water users and conservationists on one side, and the ski development and real estate interests on the other side. One of the main questions in this battleground is how much land do skiers need? This can help determine other water impacts.

MORE ACREAGE THAN SKIERS?

Since the 1978–1979 ski season, the number of U.S. skiers has not increased substantially. Ski areas and their snowmaking areas are increasing their slopes to compete for skier revenue. This generally requires more land and more water. Most ski areas feel they must now do so in order to remain economically viable. Ninety percent of ski areas in the western United States are on public lands administered by the Forest Service. When the Forest Service expands ski areas and snowmaking is used, it creates a battleground. Conservationists contend that by doing so the Forest Service encourages ski area expansions without due consideration to all the environmental impacts and for other public recreation needs such as hiking or climbing.

Construction activities for ski lodges are a significant problem. Moving large amounts of soil around the side of a mountain is inherently risky to the workers and the environment. In site preparation they scrap large areas of soil in the mountain environment. All construction equipment can leak and require on-site storage of gas, oil, antifreeze, leaded batteries, solvents, and parts. These leaks can contaminate both surface water and groundwater. Snowmaking utilizing water of lower quality, including that from treated sewer systems, can have an affect on the watershed and downstream users.

TRANSPORTATION AND AIR POLLUTION

Travel to and within ski resort areas creates transportation stresses and air quality problems. By engaging in programs within the community and region to reduce the number of vehicle miles traveled, ski areas can help mitigate these environmental problems. They can us buses, shuttles, and trains for workers and skiers, for example.

Ski areas are among the first businesses to suffer the impacts of global warming. Although the impacts are greatest at the poles, mountain snow is very sensitive to warmer temperatures. Warmer temperatures generally mean more rain, less snow. Ski lodges then have to make enough snow to keep the season open long enough to remain economically viable. Ski lodges use large amounts of energy in terms of heating and cooling, guest amenities like spas and swimming pools, and transportation. Because of their sensitivity to global warming issues, some ski lodges are investigating alternative energy forms.

INDUSTRIAL RECREATION AND VIOLENCE

The Earth Liberation Front committed the October 21, 2006, Vail, Colorado, ski resort arson, which did about $12 million in damage to four ski lifts and five buildings. The group said it struck to protect the lynx, maintaining that the 880-acre back-bowl expansion would damage lynx habitat. This arson was to attract public attention to the alleged collusion between recreation corporations and the government agencies managing federal lands.

Environmentalists claim that industrial recreation could be more damaging to wild land than all the logging, mining, and grazing of the past century. Industrial recreation refers to intensive, mechanized use of the landscape by high numbers of people. It usually involves extensive infrastructure development. Widespread recreational development will change the nature of the wilderness experience.

The major ecological problem with industrial recreation is that the federal government rarely requires ecological analysis of its effects on the environment and has never looked at the cumulative or synergistic impacts of such development. A ski area has never been required to complete an environmental impact statement (EIS) unless the Forest Service was sued to provide one. Even then, an EIS is advisory only.

FAILURES OF MITIGATION MEASURES

Ski resorts are in sensitive environments and provide a stark testing ground for mitigation measures, activities taken to lessen the environmental impacts. Mitigation is a battleground because it is often required for a project's approval but rarely enforced post–project approval. Some conservationists believe that it is not possible to mitigate the environmental impacts of a large ski development and that ski lodge projects and expansions should simply be denied.

Most projects having a significant impact on the environment have to file some type of environmental assessment or environmental impact assessment. There are many exceptions or categorical exclusions, and many question how truly rigorous these assessments are in the United States. The agency decision maker is not required to make the most environmentally sound decision but considers all kinds of political and economic factors in the final decision. One common requirement and recommendation that comes out of these assessments is that the projects mitigate their worst environmental impacts. Promises of mitigation exist in almost every final environmental impact statement, and many question their enforcement. Some mitigation measures themselves may have significant environmental impacts, and they are an exception. One measure of the mitigation is its express intent, generally to restore the environment the way it was before significant environmental impacts. By this environmental measure, most mitigation measures in the Untied States fail to meet promised expectations. This is the subject of controversy and litigation.

Many mitigation projects have failed due to one or more of the following reasons:

- poor siting and project design;
- inadequate monitoring and enforcement programs;

- lack of adequate maintenance, knowledge, or remedial activities; and
- failure of project proponents to comply with the conditions of their permits.

Based on over a decade of results, the cumulative record of past mitigation projects remains poor overall, with few examples of long-term success. Under present mitigation policies and practices, environmental losses are likely to be long lasting and mitigation as currently practiced has a high chance of failure. The weakly enforced mitigation policy adds fuel to the battleground because many environmentalists believe that the ski development should be denied.

SKI RESORTS AND SUSTAINABILITY

The environment is a ski resort's most significant asset. But skiing, snowboarding, and countless summer recreational activities often come at the cost of the environment. This past summer the National Ski Areas Association (NSAA) and its partner organizations started Sustainable Slopes: The Environmental Charter for Ski Areas to help protect the very environment upon which ski resorts depend. The charter proposed ways that participating ski resorts can manage issues like wastewater treatment facilities, stormwater runoff, and erosion and sedimentation, while demonstrating their commitment to good environmental stewardship.

The partner organizations include a host of federal, state, and local agencies such as the EPA, USDA Forest Service, the Conservation Law Foundation, National Fish and Wildlife Foundation, Leave No Trace Inc., The Mountain Institute, and the U.S. Department of Energy. The EPA supported the development of the Environmental Charter and continues to support the initiative by providing technical assistance from existing voluntary partnership programs such as the Water Alliance for Voluntary Efficiency, the Waste Wise program dealing with solid waste, the Energy Star program that promotes energy efficiency, and the agency's smart growth and development efforts.

CHARTER OVERVIEW

The Environmental Charter for Ski Areas is a voluntary initiative that holds participating ski areas (winter and summer resort operations) to a broad set of principles that provide a framework for implementing best management practices, assessing environmental performance, and setting goals for future improvement. More than 160 ski areas, representing 31 states that host 70 percent of the country's skiers and snowboarders, have already endorsed the charter. To recognize the resorts' participation, NSAA issues a Sustainable Slopes endorsement logo to each resort to display at their facility and on their marketing materials.

NSAA and the partner organizations are hoping that all resorts in the country will endorse and adopt the principles. Most of the resorts who have yet to endorse the charter are small resorts without the staffing and financial or technical ability to implement the principles. To address this problem, the partner organizations will continue to develop tools and education programs that will make it easier for all resorts to eventually endorse the principles. The voluntary principles are meant to provide overall guidance to help ski resorts practice good environmental stewardship. They are not a list of legal requirements that must be applied in every situation. Since each ski resort operates in a unique local environment or ecosystem, each resort reflects regional differences; therefore, each resort must make its own decision about how to achieve sustainable use of natural resources. While individual resorts have the same overall goal of implementing the charter, they will need to choose different paths to get there.

The Principles

The principles in the charter were developed through a collaborative effort by NSAA and the partner organizations and are intended to be updated periodically as needed. They focus on three areas.

- Planning, Design, and Construction. The principles include (1) engaging stakeholders in dialogue on development plans and implementation; (2) planning and siting facilities to avoid negative impacts on natural resources and to avoid sprawl; (3) designing new facilities to conserve water, energy, and materials; and (4) meeting or exceeding all regulatory requirements.
- Operations. The principles include (1) optimizing efficiency and effectiveness of water uses throughout the ski resort; (2) protecting and minimizing wildlife and habitat impacts; (3) maintaining minimum stream flows; (4) conserving water, energy, and fuel; (5) managing wastewater responsibly; (6) reducing all waste generated at the ski resorts; (7) reusing and recycling where possible; (8) minimizing air quality impacts; (9) designing resorts to complement the natural environment, and (10) contributing to solutions to decrease transportation issues. Several suggestions are offered for each resource. For example, to protect water quality, the charter suggests that ski resorts participate in watershed planning and management efforts, maintain vegetative buffers along streams to improve natural filtration and protect habitat, and apply appropriate stormwater management techniques and erosion and sediment control practices.
- Education and Outreach. The principles include (1) promoting environmental education and awareness and (2) enhancing the relationship between the ski area and stakeholders so that it benefits the environment. Ski resorts are expected to promote the Environmental Code of the slopes, a list designed to heighten the public's awareness about ways they can make sustainable use of natural resources while participating in outdoor recreational activities. The code suggests that the public practice energy conservation, participate in educational events and cleanup days sponsored by the ski resort, and practice outdoor ethics like respecting wildlife and not littering.

Based on information collected each year from the resorts, the NSAA will issue an annual report card to assess how well the participating resorts are meeting the goals outlined in the principles. To assist with data collection, NSAA and the partner organizations plan to work together over the next few years to set and achieve measurable goals for all the principles. NSAA expects the annual report to become more quantitative as the program develops and grows. The report will be issued each year in May and will be available on the NSAA Web site (www.nsaa.org).

Compliance Incentives

A series of incentives encourage ski resorts to adopt and follow the charter. First, by doing so, ski resorts demonstrate their environmental stewardship to customers and partner organizations. Second, adopting and implementing the principles results in a reduction of waste and energy use, which benefits the environment and resort profits. Third, partner organizations will provide technical and/or financial assistance to make improvements and share data across the industry. Resorts can also win an award, sponsored by the Skiing Company (a Times Mirror company), for environmental excellence based on the principles. And finally, by following the charter, ski resorts can ensure their livelihood is sustainable for the future through customer satisfaction and environmental protection.

Several ski resorts have developed management plans based on the charter.

Crystal Mountain, Washington: Their Master Development Plan includes several measures to address the protection of water resources, including water reclamation and conservation. Under this master plan, they have also developed management plans for roads, trails, stormwater, and stream restoration. The management plans are implemented to offset temporary and permanent watershed impacts, and monitoring is used to verify their implementation, determine the effectiveness of the restoration, and validate the maintenance of improvement of the watershed functions.

Snoqualmie Pass, Washington: The Summit at Snoqualmie's Master Development Plan will be similar to that of Crystal Lake. However, the Summit must address key wildlife corridors that represent the very heart of the northern spotted owl controversy. Any proposed expansion of facilities at the Summit must include sufficient revegetation or preservation of previously cleared forest to be determined neutral or beneficial to old-growth forest. The proposed development will be required to include up to 400 acres dedicated to old-growth forest preservation. In addition, several areas within the ski terrain will be revegetated to improve the aesthetics. This preservation/revegetation approach will benefit skiers, wildlife, and watershed function.

Westwood, California: Dyer Mountain is the only undeveloped resort to endorse the environmental charter. In 1998, Dyer Mountain Associates was formed, and plans to develop a resort community began to unfold. In 1999 several groups of consultants researched existing data and conducted field investigations to locate key habitats in environmentally sensitive areas and any other constraints to developing ski, golf, residential, and base-area facilities. Based on the vision for the recreational community and the environment, the consultants determined that four elements would drive the planning for the project, all targeted toward keeping the sense of place that currently dominates the site. These elements include emphasizing general environmental protection and using innovative planning and design to avoid environmental impact, while also offering the technology and recreational amenities desired by the community.

POTENTIAL FOR FUTURE CONTROVERSY

Ski resort development presents the current preferred recreational use for public lands. Packhorses, hikers, mountain bikers, the handicapped, dog owners, rock climbers, hang gliders, balloonists, and various sorts of aerial artists (base jumping, tightropes, etc.) are all recreational users, and each is controversial. Ski resort development represents a significant increase in environmental impacts as acreage, traffic, and year-round operations increase.

These environmental impacts take place not just anywhere, but in places many have preserved for environmental reasons. Mountain environments can be sensitive ecotones that are easily irreparably damaged. In these environments it is essential that cumulative impacts be accurately measured, and the effects of development recorded. Cumulative risk assessment methodologies have been done in Banff and British Columbia in these ecosystems. This may be an area where any promised corporate mitigation or strategy simply will not work in terms of environmental preservation. It may also be a chance to find out what works in these environments for purposes of sustainability. This is one argument used by a large, sustainable resort close to Ayers Rock, Australia, a sacred site to the indigenous people. Treaty rights of indigenous people can also be an issue for ski resort development in some places.

It is likely that controversy about the environmental impact of the size of ski resorts will continue, especially on public lands. The direction of this controversy is likely to place ski resort development in with other park users and concessionaires and their environmental impacts. Some ski resorts have embraced sustainability and may help discover practical approaches that other communities can use. Critics say these policies are voluntary and based on self-reported information that is self-serving. However, as long as ski resort development continues to have large environmental impacts, it is likely that environmental lawsuits will continue.

See also Climate Change; Cumulative Emissions, Impacts, and Risks; Endangered Species; Environmental Impact Statements: United States; Federal Environmental Land Use; Logging; Sacred Sites; Sustainability; Watershed Protection and Soil Conservation

Web Resources

National Ski Areas Association. Available at http://www.nsaa.org/. Accessed January 22, 2008.

Further Reading: Baron, Jill S. 2002. *Rocky Mountain Futures: An Ecological Perspective.* Washington, DC: Island Press; Kimmins, Hamish, and J. P. Kimmins. 1997. *Balancing Act: Environmental Issues in Forestry.* Vancouver: University of British Colombia Press; Körner, E. Christian Spehn, and Ch M. Korner. 2002. *Mountain Biodiversity: A Global*

Assessment. Oxford: Taylor and Francis; Murphy, Peter E., and Ann E. Murphy. 2004. *Strategic Management for Tourism Communities: Bridging the Gaps.* Clevedon, UK: Channel View Publications; Shabecoff, Philip. 2003. *A Fierce Green Fire: The American Environmental Movement.* Washington, DC: Island Press.

SOLAR ENERGY SUPPLY

Controversies around solar energy include economic arguments that it is not always cost-effective. Land-use restrictions can also impede solar energy applications. Proponents point to the renewable nature of this source of energy as worthy of investment now as other nonrenewable sources become depleted.

A consistent aspect of U.S. environmentalism is conservation of natural resources. When electricity comes from dams, gas, coal, nuclear plants, or other sources, the true cost can increase dramatically in a short time because of the lack of renewability of many of these resources. Many of these fossil fuel-based power sources have large environmental impacts, from the resource extraction to use to pollution and waste. Saving power is a tenet of many environmentalists. Using less power considered a softer ecological footprint. Frugal homeowners also find it a convenient way to save money. Those interested in sustainability find solar power to be acceptable. Many developing nations like solar power because that is the only electricity available. Many countries have limited hours of electricity, if any at all. Power sources can be unreliable, with frequent brownouts. Solar power combined with low-power LED lights is currently bringing light to the night in small communities in developing nations. Instead of relying on a distant grid of wires, solar power can be produced on site. This also greatly reduces its environmental impacts, or its ecological footprint.

Alternative energy sources always trigger initial controversies about cost-effectiveness. These can be complex debates including utility rate structures; bond recovery rates; local, state, and federal regulatory accommodations; safety; and scientific foundations. Most measures of cost-effectiveness measures compare the new source with petrochemical sources. Since the rate of petrochemical depletion is scientifically debated and politically contested, it remains an unknown factor in cost-effectiveness computations. Another debated assumption in these measures is the provider of the power. Solar energy can be home generated, or off the grid. In some areas the power company is required to buy back the excess power generated, provided the correct monitoring mechanisms are in place.

SOLAR ENERGY BASICS

The sun is the source of almost all of the energy on Earth. Converting sunlight to electricity or heat is *solar energy*. There are many solar technologies and technological variations with the use of solar energy. The most basic way is the use of photovoltaic systems to convert sunlight directly into electricity. These systems are commercially marketed as photovoltaic (PV) arrays. They are used to generate electricity for a single residential or commercial building. Large arrays can be combined to create a solar power plant using heat. The sunlight is focused with mirrors to create a high-intensity heat source. This heat then produces steam power to run a generator that creates electricity.

Solar water-heating systems for most residential and commercial buildings usually have two main sections. The first is a solar collector located somewhere the sun can hit it. Generally the longer and more direct the sunlight the more power created, depending on the efficiency of the PV array. The second section, connected to the solar collector, is a storage tank of liquid, generally water, which retains the heat collected by the solar collector. This heated water can then be used for heating, washing, cleaning, and so on. The sun heats the solar collector that heats the fluid running through tubes within the collector.

Solar power can be used for anything that requires electricity. The most traditional buildings use it to heat hot water, which is very energy intensive. Many residential commercial buildings can use solar collectors to do more. Solar heating systems can heat the buildings. A solar ventilation system can be used in cold climates to preheat air before it enters a building. Other than buildings, solar power is the energy source for space missions, remote viewing and sensing outposts in wilderness areas, home motion detector lights, and many, many other applications.

PASSIVE SOLAR HEATING: COOLING AND DAYLIGHTING

Structure and design with nature often mean incorporating the sun into the energy plan for the building. They are passive systems because they require little action once built. However, to efficiently use the sun as a passive energy source requires strict compliance with the rules of nature at the particular site. Buildings designed for passive solar and daylighting require design features such as large south-facing windows, building materials that absorb and slowly release the sun's heat, and structural designs to support any holding tanks. Passive solar designs should include natural ventilation for cooling. The way a building is situated on a lot can have a large effect on the efficiency of solar passive energy.

Many municipal zoning and land-use laws are strict about the zoning envelope, especially for residential structures. Buildings must be set back from the front, back, and side, creating an envelope around it. On a small lot this will greatly constrain the direction the building can face. This can limit passive solar efficiency, as well as active solar systems. This is one land-use battleground related to solar power.

ENVIRONMENTAL REQUIREMENTS

The environmental requirements for solar power differ based on power usage. Often they are site specific and not always readily available to a home buyer or builder. Essential environmental information is how much solar energy is available to a particular solar collector. The availability of or access to unobstructed sunlight for use both in passive solar designs and active systems is protected by zoning laws and ordinances in some communities.

COMMUNITY SOLAR ACCESS VERSUS PRIVATE PROPERTY

Access to the sun has always been controversial in urban settings. Access to light and air was thought to be healthier and became an important part of U.S. private property law. Solar access is the access to unobstructed, direct sunlight. Modern solar access battlegrounds emerged in the United States when commercial property owners wanted to protect their investment in solar power from nearby development casting shadows. This can be a contentious land-use battleground because it can, arguably, infringe on the development rights of nearby property owners. Advocates of solar energy point out that community-wide solar access can greatly increase the energy efficiency of the solar collectors and lower the cost of energy.

Several communities in the United States have developed solar access land-use guidelines and/or ordinances, but most have not done so. There is ongoing tension between current land-use laws and zoning ordinances and modern energy technologies that fuels this controversy about protected solar access, individually or community wide. Many communities are overwhelmed with cell tower debates and economic development. Real estate, banking, and mortgage lending interests are very important economic development stakeholders. They prefer traditional private property approaches. Many private property owners see mandatory solar access as another infringement of their rights in private property, like cable television wires, sewer pipes, and cell phone towers. Zoning is a common mechanism used to protect solar access, but can face community resistance if mandatory. This means that if you install a solar collector it may be in the shadow of another developed building later. Environmentalists would like to see more protection for investment in solar energy. Many states now offer tax rebates for alternative energy sources, including appliances, installation, and use.

CONTROVERSIES WHEN DESIGNING THE LAND-USE PLAN

Communities have created their own community solar access policies and land-use laws. When there are no local solar access laws, it is still possible to buy it from potential shadow sources. Landowners can purchase the surrounding development rights. It is possible for governmental entities to exercise their taking power to achieve public purposes related to solar energy development. Both these land-use policies stir up considerable controversy.

Traditional zoning ordinances and building codes can create problems for solar access. Most pertain to the zoning envelope mentioned previously.

- Height
- Setback from the property line
- Exterior design restrictions
- Yard projection
- Lot orientation
- Lot coverage requirements

The most important solar access regulation for land development is to face the sun in a predominantly east-west street direction. Common problems those wishing to install solar power have encountered with building codes include the following:

- Exceeding roof load
- Unacceptable heat exchangers
- Improper wiring
- Unlawful tampering with potable water supplies.

Potential zoning issues include the following:

- Obstructing side yards
- Erecting unlawful protrusions on roofs
- Siting the system too close to streets or lot boundaries.

Special area regulations such as local community, subdivision, or homeowner's association covenants also demand compliance. These covenants, historic district regulations, and floodplain provisions can easily be overlooked.

POTENTIAL FOR FUTURE CONTROVERSY

Voluntary consumer decisions to purchase electricity supplied by renewable energy sources represent a powerful market-support mechanism for renewable energy development. Beginning in the early 1990s, a small number of U.S. utilities began offering green power options to their customers. Since then, these products have become more prevalent, both from utilities and in states that have introduced competition into their retail electricity markets. Today, more than 50 percent of all U.S. consumers have an option to purchase some type of green power product from a retail electricity provider. Currently, about 600 utilities offer green power programs to customers in 34 states.

This burgeoning economic growth will push at the current constraints on the use of solar energy. As more solid information comes in about the true environmental costs of petrochemical pollution, the amount of oil left, and the large, record-breaking profits made by multinational petrochemical corporations, communities and residents seeking more self-sufficiency will pursue solar power.

See also Arctic Wildlife Refuge and Oil Drilling; Sustainability; "Takings" of Private Property under the U.S. Constitution

Web Resources

Solar Energy Links. Available at www.solarbuzz.com/Links/Environmental.htm. Accessed January 22, 2008.

World Conservation Union. Energy and Environment. Available at www.iucn.org/en/news/archive/2001_2005/press/mb_energy.pdf. Accessed January 22, 2008.

Further Reading: Gordon, Jeffrey. 2001. *Solar Energy: The State of the Art*. London: James and James/Earthscan; Laird, Frank N. 2001. *Solar Energy, Technology Policy, and Institutional Value*. Cambridge: Cambridge University Press; Oldfield, Frank. 2005. *Environmental Change: Key Issues and Alternative Perspectives*. Cambridge: Cambridge University Press; Scheer, Hermann. 2005. *A Solar Manifesto*. London: James and James/Earthscan.

SPRAWL

Sprawl is the unconstrained growth of real estate and unplanned land development. It has large environmental impacts. Controlling sprawl requires land-use regulation that can decrease the value of some property.

WHY IS SPRAWL AN ENVIRONMENTAL CONTROVERSY?

Sprawl refers to a sprawling use of natural resources, especially land. This inefficient use of natural resources and open spaces increases avoidable and unnecessary environmental impacts, contend antisprawl groups. Poorly planned real estate development threatens our environment, our health, and our quality of life in many ways. Sprawl spreads development out over large amounts of land, paving much of it. Because Americans do not live near where they work, and land-use planning tends to separate industrial, commercial, and residential land uses, there are long distances between homes, services, and employment centers. This increases dependency on the car, which pollutes more of the environment. Sprawl decreases pedestrian or bicycle transit routes and can have a negative impact on individual and public health.

Sprawling development does increase environmental impacts in air and water. As reliance on cars and pavement of more roads increases, so does air and water pollution. Sprawl destroys more than two million acres of parks, farms, and open space each year.

The owners of the parks, farms, and open space sell their property willingly on the private real estate market or unwillingly when taken by government through eminent domain processes. Sprawl can begin when rural areas are allowed to subdivide their large tracts of land into smaller parcels of land for residential development. After this counties and cities decide the minimum lot size allowed for a residential use. These residential developments, called subdivisions can be large or small, built all at once or in phases, and increase demand for government services. One primary government service they demand is roads.

Sprawling development increases traffic on side streets and highways. It pulls economic resources away from existing communities and spreads them out over sparse developments far away from the core. Local property taxes subsidize new roads, water and sewer lines, schools, and increased police and fire. An underlying concern with this controversy is that it leads to degradation of our older towns and cities and higher taxes. The relationship to taxes is itself a battleground. The underlying idea is that municipal services like fire, police, sanitation, and education all cost more because development is more spread out, or sprawling.

Suburban sprawl uses much open and green space, increases air and water pollution. Critics claim sprawl is an institutional force, supported by tax policies, land speculation, and an unrestrained profit motive. Others claim sprawl is simply the result of unrestrained market dynamics applied to land development for profit. People move to outlying areas because land is cheaper, they reason.

SPRAWL-THREATENED CITIES

U.S. cities are suffering from sprawl. In the United States, many municipalities may make up a given metropolitan area. They can compete with one another to develop a high tax base with low service delivery. They seldom act in a regional manner with the exception of some transportation planning. As a result, development occurs in an unplanned manner. What are the actual descriptions of sprawl? What are the environmental impacts? How do sprawl controversies unfold?

Washington, D.C.

The District of Columbia has steadily lost population since 1970. The outermost suburbs have experienced growth. Open space is being rapidly allocated to commercial and residential structures, roads, parking lots, and strip malls. A 1994 federal study ranked Washington, D.C., number one (ahead of Los Angeles) in the cost per person of wasted fuels and time spent stuck in traffic jams.

Cincinnati, Ohio

While the number of people moving into the Cincinnati metro area has not risen significantly in recent years (8 percent in the 1980s and 2.2 percent from 1990 to 1996), its land area has spread out steadily over the years: from 335 square miles in 1970 to 512 square miles in 1990, a 53 percent increase. The area grew by another 12 percent between 1990 and 1996. The average number of daily vehicle miles traveled per person increased by 29 percent between 1990 and 1996.

Kansas City, Missouri

The metro area has also been influenced by an extensive regional freeway system planned in the 1940s and white flight. Kansas City has paved miles of roads, sidewalks, curbs, and even streambeds. Kansas City has more freeway lane miles per capita than any other city in the country. The percentage of work trips made by people driving alone is 79.7 percent, above the national average of 73.2 percent. Public transit is poor, Public transit ridership per capita in Kansas City is one-third the average of most other cities the same size.

Seattle, Washington

The Seattle metropolitan region is moving southward along the coast and eastward, closer to the Cascades mountain range. The metropolitan area grew in population by 13 percent from 1990 to 1996, much of it in the outer suburbs. During the same period, population grew by only 1.6 percent in Seattle's center city. Seattle's four-year-old urban growth boundary has helped slow down some of the unplanned sprawl.

Minneapolis-St. Paul, Minnesota

Between 1982 and 1992, Minnesota lost 2.3 million acres of farmland to development. Hennepin County, where Minneapolis is located, lost the greatest proportion: 29 percent. The rate of open space destroyed by development increased by almost 25 percent in the Minneapolis-St. Paul metro area overall. The number of people moving to the city's surrounding areas increased 25 percent in the 1980s and another 16 percent in the early 1990s. Few urban areas have experienced a faster-growing traffic problem than Minneapolis/St. Paul.

Sprawl is a development pattern that affects all sizes of cities. It can have the same effects and controversies as in large cities.

FIVE MOST SPRAWL-THREATENED MEDIUM CITIES (POPULATION: 500,000–1 MILLION)

1. Orlando, FL
2. Austin, TX
3. Las Vegas, NV
4. West Palm Beach, FL
5. Akron, OH

FIVE MOST SPRAWL-THREATENED SMALL CITIES (POPULATION: 200,000–500,000)

1. McAllen, TX
2. Raleigh, NC
3. Pensacola, FL
4. Daytona Beach, FL
5. Little Rock, AR

ALTERNATIVES TO SPRAWL

Hundreds of urban, suburban, and rural neighborhoods are using smart-growth solutions to address the problems caused by sprawl. Examples of smart-growth solutions include:

- More public transportation
- Planning pedestrian-friendly developments with transportation options; providing walking and bicycling facilities around services and parks
- Building more affordable housing close to transit and jobs
- Requiring greater public involvement in the transportation and land-use planning processes
- Requiring developers to pay impact fees to cover the costs of new roads, schools, and water and sewer lines, and requiring environmental impact studies on new developments

SMART GROWTH

In response to sprawl, a movement to emphasize planning in land development exists under the name of smart growth. Smart growth is development that serves the economy, the community, and the environment. It changes the terms of the development debate away from the traditional growth/no growth divide to "how and where new development should be accommodated." Underscoring the smart-growth movement is the premise of preserving open space, farmland, wild areas, and parks as necessary for a healthy environment and community. The question of how to stop sprawl is complicated, new in the United States, and controversial. Smart growth has garnered the support of some state legislatures. However, any loss of profit in the sale of private property due to the local land-use rules required by smart growth will encounter stiff resistance from powerful lobbies of realtors, home builders, mortgage bankers, and others with a financial interest in land.

Ecosystem Preservation

Given the current U.S. checkerboard pattern of many competing municipalities in any given metropolitan area, any shift toward ecosystem preservation will be extremely difficult, but many claim it is necessary. The shift is prompted by the realization that ecosystems are the appropriate units of environmental analysis and management. Wildlife must be managed as a community of interrelated species; actions that affect one species affect others. The open space plan emphasizes connections to off-site habitat and preservation of corridors rather than isolated patches. It helps to preserve patches of high-quality habitat, as large and circular as possible, feathered at the edges, and connected by wildlife corridors.

Patches preserved in an urbanizing landscape should be as large as possible. In general, the bigger the size of land, the more biodiversity of species it can accommodate. Patches of 15 to 75 acres can support many bird species, smaller mammals, and most reptiles and amphibians. Wildlife corridors should be preserved to serve as land bridges between habitat islands. Riparian strips along rivers and streams are the most valuable of all corridors, used by nearly 70 percent of all species.

When land is developed, a large volume of stormwater that once seeped into the ground or nourished vegetation is deflected by rooftops, roads, parking lots, and other impervious surfaces; it ends up as runoff, picking up urban pollutants as it goes. This change in hydrology creates four related problems. Peak discharges, pollutant loads, and volumes of runoff leaving a site increase, as compared to predevelopment levels. By reducing groundwater recharge, land development also reduces base flows in nearby rivers and streams. To mitigate the adverse impacts of development, there are two options: stormwater infiltration and stormwater detention. With infiltration, stormwater is retained on-site in basins, trenches, or recharge beds under pavements, allowing it to infiltrate into the ground. With detention, stormwater runoff is slowed via swales, ponds, or wetlands but ultimately discharged from the site. Experts are beginning to favor infiltration as the only complete approach to stormwater management. Where soils and water table elevations permit, infiltration can maintain the water balance in a basin and runoff before and after development using infiltration trenches, swales, different dams, and/or permeable pavements. Infiltration rates can be increased by means of infiltration basins and vegetated swales, created prairies, created wetlands, and a stormwater lake to reduce runoff volumes. The swales and prairie lands clean and infiltrate runoff, while the wetlands and lake polish the outfall. Turf is used only where it serves a specific purpose, such as erosion control or recreation, rather than as fill-in material between other landscape elements. One visual preference survey found that lawns with up to 50 percent native groundcover are perceived as more attractive and less work (as well as much more natural) than are conventional turf lawns. Plants with similar irrigation requirements are grouped together into water-use zones (so-called hydrozones). Irrigation systems can then be tailored to different zones rather than operating uniformly. It is recommended that high water-use zones (consisting of turfgrasses and plants that require supplemental watering year-round) be limited to 50 percent of total landscaped area, and that drip or bubbler irrigation be used on trees, shrubs, and ornamentals. Even some of the most manicured developments are beginning to experiment with native plantings. Expect to see more of the same as other developers discover that a palette of native and adapted plants is more economical and visually pleasing than is endless turfgrass.

The required environmental changes in our approaches to sprawl are severe in our current context. Many are nonetheless required. This assures that as they become more operational in education, business, and governmental practices they will also be controversial.

POTENTIAL FOR FUTURE CONTROVERSY

The land-use decisions made today could have the most important, long-term environmental consequences for future sustainability. Innovative thinking and foresight can facilitate the creation of greenspace in development plans and how urban communities can create greenspace from previously ignored areas. The vast majority of land is privately owned. As a result, individual landowners, developers, and local governments are the principal land-use decision makers. They do not always have the same vision and foresight regarding the environment if it affects their profit from the sale of the land.

U.S. metropolitan areas are spreading outward at unprecedented rates, causing alarm from Florida to California, from New Jersey to Washington State. Without changes in policy and practice, most new development will take the form of suburban sprawl, sprawl being this nation's now-dominant development pattern. The economic and social costs will be large. By designing with nature, developers can further the goals of habitat protection, stormwater management, water conservation, and aquifer protection. Ways of furthering another environmental goal—air quality—can include natural amenities such as woodlands, hedgerows, slopes, rock outcroppings, and water bodies, which cost nothing in their pure state and are preferred by residents. Wild places (natural areas with nothing done to them at all) are a particular favorite with children. Greenbelts and other open spaces, if designed for physical and visual access, can enhance property values of nearby developable lands.

With increasing population and a strong, car-based transportation infrastructure, sprawl will continue. But strong environmental and public health values oppose the negative impacts of sprawl. With a long tradition of respecting private property but with a need more brass-tacks environmental policies, controversies will continue to develop.

See also Air Pollution; Ecosystem Risk Assessment; Land-Use Planning in the United States; Sustainability; "Takings" of Private Property under the U.S. Constitution; Transportation and the Environment; Water Pollution

Web Resources

Sierra Club. Stopping Sprawl. Available at www.sierraclub.org/sprawl/. Accessed January 22, 2008.
Sprawl Watch. Available at www.sprawlwatch.org/. Accessed January 22, 2008.
U.S. Environmental Protection Agency. Smartgrowth. Available at www.epa.gov/smart growth/about_sg.htm. Accessed January 22, 2008.

Further Reading: Garreau, Joel. 1991. *Edge City: Life on the New Frontier.* New York: Random House; Hayden, Dolores. 2004. *A Field Guide to Sprawl.* New York: W. W. Norton and Co.; Sierra Club. 1999. *Solving Sprawl: The Sierra Club Rates the States.* Washington, DC: Sierra Club; Wolch, Jennifer, Manual Pastor, and Peter Dreier, eds. 2004. *Up Against the Sprawl.* Minneapolis: University of Minnesota Press.

STATE ENVIRONMENTAL LAND USE

States exert strong control over their public lands. As such they face some of the same controversies of multiple uses on land meant for conservation as does the federal government. Additionally state environmental agencies are delegated authority over many environmental laws via memorandums of understanding from regional offices of federal agencies like the U.S. Environmental Protection Agency. Millions of dollars pour into states via these memorandums about environment laws, especially enforcement by the state environmental agency of federal environmental laws. State agencies are often delegated the authority to issue industrial permits. Many state environmental agencies employ more permit writers to assist industry with compliance in single consultations then they do environmental enforcement personnel. Controversies about environmentally degrading uses on state-owned public lands (like parks), enforcement of environmental laws, permit renewal and modification, and new permits all evoke considerable consternation from communities and local elected officials.

BACKGROUND OF LEGAL BASIS FOR STATE CONTROL

As individual states became more urbanized, land development and use regulations became acceptable methods for resolving conflicts. Metropolitan areas, such as New York, Boston, Cleveland, and Los Angeles, were the first to face the land-use controversies that accompanied urban growth. At that time, local authority over land-use control was conducted on a case-by-case and piecemeal basis. The result was a hodgepodge of developmental regulatory policies, programs, ordinances, and resolutions that numbed the mind and destroyed coherent approaches to community planning. This confusing situation led to the

Standard State Zoning Enabling Act (SSZEA). This has been adopted in some form by most states.

The U.S. Constitution defines the federal government's relationship to the states, but the Constitution does not refer to local governments because local governments are considered to be creatures of state government. Thus, they derive their authority from the laws of the states where they are located, and they possess only those powers given to them by the state, either through state constitutions or state statutes, or both. The power of localities to control the uses of land within their borders derives from state statutes and, like many other areas of the law, states have not uniformly granted broad authorization for local governments to engage in land-use planning and zoning. There are a number of recent trends worth noting.

In 1999, the American Planning Association (APA), as part of its multiyear Growing Smart effort, surveyed state laws on local land-use planning to determine how many states continue to authorize planning based on the 1928 Standard City Planning Enabling Act. The survey found that almost half of the states (24) had not updated their local planning statutes since 1928, and only 11 states had adopted substantial updates of their laws. Seven states had slightly updated their planning enabling acts, and eight states were classified as having made moderate updates. Other findings from this survey are discussed further on in the section on comprehensive land-use planning.

The long history of state and local roles in land-use planning and zoning has been an important influence on current opportunities for reforms to address environmental issues. The nation was clearly in a different place in the 1920s when the cities were grappling with myriad social and environmental stresses. But today, even with the technological revolution, newer and perhaps even more complex social and environmental issues are being confronted as the nation strives to achieve some level of sustainable development and as the economic, environmental, and equity challenges are no longer contained within cities but now spread throughout suburban and rural communities as well. The American Planning Association has identified many factors to consider in reforming state planning statutes, including:

- Ongoing problems of housing affordability,
- Lack of housing diversity,
- Exposure of life and property to natural hazards, and
- The obligation to promote social equity—"the expansion of opportunities for betterment, creating more choices for those who have few"—in the face of economic and spatial separation.

Various planning and zoning enabling statutes have had an impact on the ability of local governments to consider and address environmental concerns by controlling land use. There is a renewed interest in modernizing and reforming many states' outmoded planning and zoning laws. This interest presents a unique opportunity for environmental and community advocates to provide leadership by securing the passage of revised state enabling statutes that will empower local governments to address these issues more effectively through land-use planning and zoning. However, the gap between environmental planning and land-use planning at the local level is large. That is one reason the states fill the void with their version of environmental planning.

POTENTIAL FOR FUTURE CONTROVERSY

Environmental planning is the analysis of how people impact natural resources. Decisions made regarding development and conservation, zoning, and land use may have visible and invisible impacts on the environment. Visible impacts include major land disturbances, loss of natural areas and wildlife, or pollution. Invisible impacts may be contamination of land, water, and air by point or nonpoint source pollution.

Environmental planning projects involve protecting natural resources for future generations, identifying problem areas (such as stormwater, erosion hazards, and threatened sensitive areas), and developing strategies for correcting those problems. Some states do this, but very few localities do so. State approaches are more focused on industrial compliance measures than on prioritizing environmental assessment and monitoring with adequate resources.

Many environmental controversies have the involvement of the states, either as property owners, caretakers of natural resources, recipients of federal environmental funds, or enforcers of environmental laws. All these controversies are complicated by the controversies tied in with the power to take private property that the state can exercise. Communities distrust state agencies that seem to cater to industry by granting permits without community input or approval. There are very few environmental policies in the United States that do not rely on the states in some way. It is likely this controversy will play out in the battleground of electoral politics. As environmental impacts become known among residents, and as the state may sometime preempt local land-use resistance, citizen frustration may result in a stronger environmental policy role for local government.

See also Federal Environmental Land Use; "Takings" of Private Property under the U.S. Constitution

Web Resources

North Carolina Division of Water Quality. Water Quality Section: State Environmental Policy Act (SEPA) Program. Available at h2o.enr.state.nc.us/sepa/index.htm. Accessed January 22, 2008.

State Environmental Planning Information. Available at www.nepa.gov/nepa/regs/states/states.cfm. Accessed January 22, 2008.

Further Reading: Environmental Law Institute. 2003. *Planning for Biodiversity: Authorities in State Land Use Laws.* Washington, DC: Environmental Law Institute; May, Peter J. 1996. *Environmental Management and Governance: Intergovernmental Approaches to Hazards and Sustainability.* London: Routledge.

STOCK GRAZING AND THE ENVIRONMENT

The grazing of cattle, sheep, and goats provides food but has environmental impacts. In some environments stock grazing can be destructive. Stockmen use 70 percent of the U.S. West for raising livestock, and most of this land is owned by the public. Experts and environmental activists consider ranching the rural West's most harmful environmental influence.

Many animals naturally graze or eat plants such as grasses and leaves. Some animals and plants develop strong symbiotic relationships with each other in the natural environment. Grazing animals in nature can fill important parts of the food chain in a given ecosystem. Many predators rely on them for food. Humans learned that raising your own animals was easier and more reliable than hunting them. Since early civilization, humans have grazed animals such as sheep, cattle, and goats. With the advent of large moving herds of the same grazing animal, environmental impacts increased, especially over time and in the context of increasing human population. Increasing population and expanding development reduce the amount of pastureland available and can increase the environmental impacts on the pastureland left. Increasing population also increases demand for food. The demand for meat and animal products drives the overall production of meat and the need for efficient industrial production processes. Part of these more efficient processes is producing the most meat per acre, which may have environmental impacts. Producing meat from pastures generally requires a minimum pasture size, depending on pasture quality and grazing animal.

GLOBAL CONTEXT

Not all pastureland is affected by large stock grazing systems. Approximately 60 percent of the world's pastureland is covered by grazing systems. This is just less than half the world's usable surface. The grazing land supports about 360 million cattle, and over 600 million sheep and goats. Grazing lands supply about 9 percent of the world's production of beef and about 30 percent of the world's production of sheep and goat meat. For an estimated 200 million people grazing livestock is the only source of livelihood. For many other people grazing animals is the basis of a subsistence lifestyle and culture.

U.S. CONTEXT

Ranching is big business in the United States. Although concentrated in the western United States, other states have some ranching interests. Hawaii, for instance, has the biggest ranch in the United States in the Parker Ranch on Hilo. One battleground is the use of federal land for grazing. The federal government is a large landowner is western ranching states. In the western United States, 80 percent of federal land and 70 percent of all land is used for livestock grazing. The federal government grants permits to ranchers for their herds to use federal lands. The mean amount of land allotted per western grazing permittee is 11,818 acres. Many ranchers own both private property and permits from the federal government for ranching public land. The public lands portion is usually many times larger than the private.

Cattle and sheep have always comprised the vast majority of livestock on public land. Cattle consume about 96 percent of the estimated total grazed forage on public land in the United States. There are some small public lands ranchers, but corporate ranchers and large individual operators control the market now. This is also a battleground because some say they are doing so with the aid of the U.S. government. Many of the permits are with long-term leases at below-market rates. Forty percent of federal grazing is controlled by 3 percent of permittees. On the national scale, nearly 80 percent of all beef processing is controlled by only three agricultural conglomerates.

ENVIRONMENTAL CHALLENGES AND BENEFITS

Stock grazing can damage the environment by overgrazing, soil degradation, water contamination, and deforestation. Seventy-three percent of the world's 4.5 billion hectares of pasture is moderately or severely degraded already. Livestock and their need for safe pastures is one reason for deforestation of tropical rain forests.

Prolonged heavy grazing contributes to species extinction and the subsequent dominance by other plants, which may not be suitable for grazing. Other wild grazing animals are also affected by the loss of plant biodiversity. Such loss of plant and animal biodiversity can have severe environmental impacts. In sensitive environments, such as alpine and reclaimed desert environments, the impacts of overgrazing can be irreversible. Livestock overgrazing has ecological impacts on soil and water systems. Overgrazing causes soil compaction and erosion and can dramatically increase sensitivity to drought, landslides, and mudslides.

Actions to mitigate environmental impacts of overgrazing include preservation of riparian areas, place-sensitive grazing rotations, and excluding ranchers from public lands. All these are battlegrounds. Ranchers resent the government telling them how to run their business and resist taking these steps because they cost them money. Excluding ranchers dramatically increases the intensity of the battleground, but is the preferred solution for many conservationists. Some of the areas that benefit from these types of mitigation include:

- grasslands, grassy woodlands, and forests on infertile, shallow, or skeletal soils;
- grassy woodlands and forests in which trees constrain grass biomass levels and prevent dominant grasses from outcompeting smaller herbs; and
- other ecosystems on unproductive soils that occur amongst grassy ecosystems within managed areas.

POTENTIAL FOR FUTURE CONTROVERSY

Given climate change, population growth, and the dependence of people on grazing animals it is likely this controversy will increase. In the United States, perceptions of a vested property right in U.S. land by ranchers, their families, and their communities clash with the reality that this is land held in trust for all citizens of the United States. As environmental restrictions on grazing on public and private lands challenge this perception, courts and federal agencies will be front-and-center battlegrounds in this controversy. As concern about endangered species and sustainability rises, so too will these battlegrounds enter this controversy.

See also Climate Change; Endangered Species; Federal Environmental Land Use; Global Warming; Rain Forests; Sustainability; True Cost Pricing in Environmental Economics

Web Resources

Holechek, Jerry L., and Karl Hess, Jr. Government Policy Influences on Rangeland Conditions in the United States: A Case Example. *Environmental Monitoring and Assessment.* Available at www.springerlink.com/content/w35t2181714478v1/. Accessed January 22, 2008.
van den Brink, Rogier, Glen Thomas, Hans Binswanger, John Bruce, and Frank Byamugisha. Consensus, Confusion, and Controversy: Selected Land Reform Issues in Sub-Saharan Africa. World Bank Working Paper, no 71. Available at www.landcoalition.org/pdf/wb06_wp71.pdf. Accessed January 22, 2008.

Further Reading: Davis, Charles E. 2000. *Western Public Lands and Environmental Politics.* Boulder, CO: Westview Press; Robbins, William G., and William Cronon. 2004. *Landscapes of Conflict: The Oregon Story, 1940–2000.* Seattle: University of Washington Press; Wilkinson, Charles F. 1993. *Crossing the Next Meridian.* Washington, DC: Island Press.

SUPPLEMENTAL ENVIRONMENTAL PROJECTS

Supplemental environmental projects (SEPs) are performed by environmental violators as alternatives to fines; they may be viewed as a form of community service (performed instead of the jail time or a monetary penalty). Conventional environmental enforcement results in fines paid to government treasuries and, often, injunctive relief (judicially imposed requirements to return to compliance). Supplemental environmental projects, on the other hand, target environmental improvement beyond the immediate violation and its effects. They are used in settlements between regulators and violators and also in settlement of private citizen suits enforcing environmental laws.

Although relatively new and very constrained by law, communities and some environmental groups advocate for their use, although some community groups are unaware of SEPs. Government environmental agencies have been lukewarm to the idea, and industry seems to be adopting a wait-and-see approach. The current enforcement of environmental law is itself controversial. Generally the Environmental Protection Agency or state environmental agencies enforce the terms and conditions of permits, licenses, and court cases. If the industry or municipality or other entity accused of environmental wrongdoing challenges it, they must first exhaust all administrative remedies before getting a court to accept jurisdiction. If a big case is headed to court, the EPA will give it to the U.S. Department of Justice for prosecution. Some have accused the EPA of being captive to industry because of their leniency toward wrongdoers.

Supplemental environmental projects are environmentally beneficial projects that a violator voluntarily agrees to undertake during settlement of an enforcement action. The purpose of an SEP is to secure significant environmental or public health protection improvements beyond those achieved by bringing the violator into compliance. The violator is not under a preexisting legal requirement to do the project. The proposed cash penalty may be lowered if the violator chooses to perform an acceptable SEP. An acceptable SEP must improve, protect, or reduce risks to public health or the environment and have a relationship with the violation. The EPA does not manage or control the money for the project. The EPA does provide enough oversight to ensure that the company follows through on what it promises to do.

CHARACTERISTICS OF SEPs

Because SEPs are part of an enforcement settlement they must meet certain legal requirements. There must be a relationship between the underlying violation and the human health or environmental benefits that will result from the SEP. An SEP must improve, protect, or reduce risks to public health or the environment, although in some cases an SEP may, as a secondary matter, also provide the violator with certain benefits.

The SEP must be undertaken in settlement of an enforcement action as a project that the violator is not otherwise legally required to perform.

SEP GUIDELINES

There are several guidelines that an SEP must meet. A project cannot be inconsistent with any provision of the underlying statute(s). An SEP must advance at least one of the objectives of the environmental statute that is the basis of the enforcement action. The EPA must not play any role in managing or controlling funds used to perform an SEP.

CATEGORIES OF ACCEPTABLE SEPs

The EPA has set out eight categories of projects that can be acceptable SEPs. To qualify, an SEP must fit into at least one of the following categories.

- Public Health: SEPs may include examining residents in a community to determine if anyone has experienced any health problems because of the company's violations.
- Pollution Prevention: These SEPs involve changes so that the company no longer generates some form of pollution. For example, a company may make its operation more efficient so that it avoids making a hazardous waste along with its product.

- Pollution Reduction: These SEPs reduce the amount and/or danger presented by some form of pollution, often by providing better treatment and disposal of the pollutant.
- Environmental Restoration and Protection: These SEPs improve the condition of the land, air, or water in the area damaged by the violation. For example, by purchasing land or developing conservation programs for the land, a company could protect a source of drinking water.
- Emergency Planning and Preparedness: These projects provide assistance to a responsible state or local emergency response or planning entity to enable these organizations to fulfill their obligations under the Emergency Planning and Community Right-to-Know Act (EPCRA.) Such assistance may include the purchase of computers and/or software, communication systems, chemical emission detection and inactivation equipment, HAZMAT equipment, or training. Cash donations to local or state emergency response organizations are not acceptable SEPs.
- Assessments and Audits: A violating company may agree to examine its operations to determine if it is causing any other pollution problems or can run its operations better to avoid violations in the future. These audits go well beyond standard business practice.
- Environmental Compliance Promotion: These are SEPs in which an alleged violator provides training or technical support to other members of the regulated community to achieve, or go beyond, compliance with applicable environmental requirements. For example, the violator may train other companies on how to comply with the law.

Other Types of Projects

Other acceptable SEPs would be those that have environmental merit but do not fit within the categories listed previously. These types of projects must be fully consistent with all other provisions of the SEP policy and be approved by the EPA.

Some have advocated for the expanded use of SEPs to preconviction phases of enforcement policy. That is, before an environmental wrongdoer or violator is found guilty they could opt for an SEP, without an admission of guilt. An admission of guilt at this stage of enforcement is very expensive for the industry in terms of future agency scrutiny, higher fines if the act is repeated, loss of community goodwill, and sometimes higher insurance premiums. A judicially supported agency finding of environmental violations or crimes will also make it easier for later courts to entertain environmental lawsuits. The industrial reputation of fighting and losing charges of pollution develops a body of evidence, and sometimes proof, that enables later suits to sustain industry's legal attempts to dismiss the case. The manufacturer of Lucite did such an SEP, remedying the water pollution caused by its emissions and empowering the community to monitor water quality and organize around environmental issues.

CATEGORIES AND EXAMPLES OF SEPs

Public Health

SEPs may include examining residents in a community to determine if anyone has experienced any health problems because of the company's violations.

City of Timpson Electric Utility

Identify all existing oil-filled electrical equipment within City of Timpson electrical utility system that contains polychlorinated biphenyls (PCBs) at one part per million (ppm) or greater. Remove from service and properly dispose of all PCBs and PCB equipment that contains PCBs at 50 ppm or greater.

Cost = $58,885.00

Pollution Prevention

SEPs involve changes so that the company no longer produces some form of pollution. For example, a company may make its operation more efficient so that it avoids making a hazardous waste along with its product.

Formosa Plastics Corporation

Replace two ethylene dichloride (EDC) furnaces seven to eight years before the end of their expected useful life.

Cost = $6,600,000.00

Pollution Reduction

SEPs are like pollution prevention, projects. Instead of eliminating the source of pollution, these projects reduce the amount and/or danger presented by some form of pollution, often by providing better treatment and disposal of the pollutant.

Micro Chemical Company

Sponsor household hazardous waste collection (HHW) event for Franklin Parish, Louisiana, for the next five years. Provide public with educational material concerning management of HHW at home. Several tons of waste will be kept out of landfills

Cost = $25,280.00

Environmental Restoration and Protection

SEPs that improve the condition of the land, air, or water in the area damaged by the violation. For example, by purchasing land or developing conservation programs for the land, a company could protect a source of drinking water.

Lead Products Company, Inc.

Remove contamination from soil and pave road to reduce exposure; provide public education and outreach.

Cost = $5,000.00

Emergency Planning and Preparedness

SEPs provide technical assistance and training to state or local emergency planning and response organizations to help them better respond to chemical emergencies.

Chem Service, Inc.

Donate $2,000 in equipment, $2,000 to Oklahoma local emergency planning committee's (LEPC) regional conference, and $4,000 in support over two years to LEPC.

Cost = $8,000.00

Assessments and Audits

Company identifies opportunities to reduce emissions and improve environmental performance.

Camp Stanley Storage Activity

Hire an on-site expert to audit and implement actions to review all inadequate hazardous waste management practices.

Cost = $555,000.00

Environmental Compliance Promotion

Provides training or technical support to other members of the regulated community to help achieve and maintain compliance, or to reduce pollutants beyond legal requirements.

National Tank Company, Inc.

Conduct Resource Conservation and Recovery Act (RCRA) and Emergency Planning and Community Right-to-Know Act (EPCRA) training seminar for local industry. Affected industry is primarily metal fabrication and coating.

Cost = $9,861.00

Other Types of Projects

SEPs that have environmental merit but do not fit within the categories listed previously. These projects must be fully consistent with all other provisions of the SEP policy and be approved by the EPA.

WANGARI MAATHAI

Wangari Maathai is the first African woman and first environmentalist to win the Nobel Peace Prize. Awarded to her in 2004, it recognizes outstanding accomplishment.

She was born in 1940 in Nyeri, Kenya. She is the inspirational founder of the Green Belt Movement. This movement uses networks of rural women to plant trees. Since 1977 they have planted over 40 million trees. She began her tree-planting citizen initiative in 1976. From there, Professor Maathai developed it into the Green Belt Movement. In 1986 that Green Belt Movement established a Pan African Green Belt Network. Several African countries have started similar environmental conservation efforts. These include Tanzania, Uganda, Malawi, Lesotho, Ethiopia, and Zimbabwe. In Africa, women are responsible for collecting firewood. Increasing deforestation has meant that women have had to travel further to get the firewood. This in turn has led to women spending less time around the home, earning income, and raising their families. In deforested areas streams often dry up faster, with no trees to shade rivers and to retain water. Women are also responsible for water collection and must also go farther for water. Sometimes the search for wood and water is risky, and it is always time consuming. By having wood and water closer to home and earning income from sustainably harvesting the trees, women became empowered through conservation.

Professor Maathai was the first woman in East and Central Africa to earn a doctoral degree. Ms. Maathai first took a degree in biological sciences from Mount St. Scholastica College in Atchison, Kansas, in 1964. She next earned a master of science degree from the University of Pittsburgh in 1966. She earned her PhD in 1971 from the University of Nairobi. She also taught veterinary anatomy there, eventually becoming chair of the department.

Professor Maathai is internationally recognized for her environmental efforts and expertise. She addressed the United Nations (UN) on several occasions. She has advocated for both the environment and for women. She has also participated in many of the most important and influential environmental summits that result in environmental agreements and treaties. She has worked with UN Commission for Global Governance and the Commission on the Future.

For further information see Wangari Maathai, *Unbowed: A Memoir* (New York: Anchor Books, 2007).

Coco Resources, Inc.

For RCRA violation, install fence around facility to separate facility from fireworks stand and day care; construct secondary containment for storage of products that contain hazardous constituents; install ventilation vents inside warehouse; and implement risk management plan. For Comprehensive Environmental Response, Compensation, and Liability Act (CERCLA or Superfund) violation, donate $500 in emergency equipment to LEPC; donate $500 to annual regional LEPC conference; and attend LEPC meetings for two years.

Cost = $11,650.00

POTENTIAL FOR FUTURE CONTROVERSY

The current legal limitations on SEPs and the fact that they are used postconviction make it difficult to judge how controversial they would be in a broader policy application. In addition, the settlements are not nearly as publicized as the efforts going into creating environmental standards, so SEPs are a form of stealth policy. By directing the resources of the wrongdoer towards benefiting the environment rather than enriching the treasury, environmental outcomes and community satisfaction could be expected to be better. However, some have expressed concern that the receipt of SEP project funds by community groups could imperil their independence and impartiality. Also, community groups wonder whether some companies perform projects as SEPs that they would otherwise be doing for business reasons.

SEPs are being used more by state and federal agencies. SEPs do rely on an environmental knowledge base that is incomplete at the present, although rapidly growing. Industries that are rooted in the community often see SEPs as a way to enhance their image. Industries that can move may not share that perception. Corporations may believe that the costs of administering SEPs outweigh the benefits that they reap in goodwill, also. Environmental groups, especially land preservation groups, embrace SEPs. But SEPs must achieve a higher profile and win over institutional resistance toward their use, before they can make substantial contributions to environmental and public health.

See also Litigation of Environmental Disputes

Web Resources

Model SEP CAFO. Available at www.epa.gov/compliance/resources/policies/civil/seps/sep cafo.pdf. Accessed January 22, 2008.

University of California, Hastings College of the Law, The Center for State and Local Government Law. "Supplemental Environmental Projects: A Fifty State Survey with Model Practices." Available at www.uchastings.edu/cslgl/SEPs.html. Accessed January 22, 2008.

U.S. Environmental Protection Agency. Call for SEP Ideas. Available at www.epa.gov/compliance/resources/publications/civil/programs/call-sepsprojects.pdf. Accessed January 22, 2008.

U.S. Environmental Protection Agency. SEP Policy memo. Available at www.epa.gov/compliance/resources/policies/civil/seps/fnlsup-hermn-mem.pdf. Accessed January 22, 2008.

U.S. Environmental Protection Agency. Supplemental Environmental Projects Library. Available at yosemite1.epa.gov/r6/r6w3c2.nsf/SEPFacility?OpenView&Start=1&Count=150. Accessed January 22, 2008.

Further Reading: Friedman, Frank B. 2003. *Practical Guide to Environmental Management.* Washington, DC: Environmental Law Institute; Payne, Scott M. 1998. *Strategies for Accelerating Cleanup at Toxic Waste Sites: Fast-Tracking Environmental Actions and Decision Making.* Singapore: CRC Press; Rechtschaffen, Clifford, and David L. Markell. 2003. *Reinventing Environmental Enforcement and the State/Federal Relationship.* Washington, DC: Environmental Law Institute; Thomas Jr., William L., Bertram C. Frey, and Fern Fleischer Daves. 2000. *Crafting Superior Environmental Enforcement Solutions.* Washington, DC: Environmental Law Institute.

SUSTAINABILITY

Sustainable development is development that places equal emphasis on environment, economics, and equity rather than economic interests above all others. It is controversial because it limits human activities in light of their environmental and equitable impacts on all affected communities.

Sustainability is controversial in the first instance because of confusion over the application of the term. Its perhaps most intuitive meaning is one that allows for the environment, or nature, to thrive. However, Western nations have used the term to mean sustainable development. Sustainable development assumes continuous economic growth, without irreparably or irreversibly damaging the environment. Human population growth is difficult for this model because it is difficult to place an economic value on the lives that exist in the future. Some environmentalists challenge the assumption of growth at all. The fundamental battleground for this emerging controversy is one of values. The continued prioritization of economic growth over environmental protection, combined with population increases, may have irreparable impacts on the environment and therefore any type of sustainable policy. So most environmentalists do use the term sustainable development because that is better than nothing at all. The underlying value differences still exist. Some values may simply not be sustainable in any sense of the term, but are important histories of powerful countries.

GLOBAL BACKGROUND

In 1987, Gro Harlem Brundtland, then-prime minister of Norway, authored a report for the World Commission on Environment and Development called "Our Common Future." In it, she described a concept of sustainable development as "development that meets the needs of the present without compromising the ability of future generations to meet their own needs." This has become the defining statement about sustainable human development. Sustainability focuses on fairness to future generations by ensuring that the ecosystems on which all life depends are not lost or degraded, and poverty is eradicated. Sustainable development seeks these goals of environmental protection and ending poverty by implementing several key concepts in development policies and practices.

The UN Conferences on Environment and Development (UNCEDs) have become the forums in which these key concepts have been turned into implementable policy statements. The agreements and statements resulting from these conferences are often identified by their host city. Perhaps the most famous of these conferences was the Earth Summit held in Rio de Janeiro in 1992. At this conference the nations of the world, including the United States, agreed to implement seven key concepts to ensure sustainable development in a declaration called the Rio Declaration and they also wrote out a work plan called Agenda 21, which remains the source of much international controversy to this day. The seven key principles emerging from the Earth Summit, and found in the Rio Declaration and Agenda 21, are:

1. Integrated decision making (three Es: environment, social equity, and economics)
2. Polluter pays
3. Sustainable consumption and production
4. Precautionary principle
5. Intergenerational equity
6. Public participation
7. Differentiated responsibilities and leadership

Another famous conference based upon this same UNCED plan was held in Kyoto, Japan, in 1997, resulting in the Kyoto protocols on climate change and the limits on emissions of greenhouse gases. Although a signatory to the protocols, the United States has not moved forward with ratification of the protocols even though it is the world's largest producer of carbon dioxide, because it disagrees with the exemptions given to developing economies like China and India.

The idea of sustainable development is a fundamentally different model of growth and development from the model of colonial and industrial development. It stresses equality in the distribution of benefits from development between developed and under- or undeveloped nations. It also stresses equal distribution of the benefits of development between contemporaries within developing nations. This egalitarian model of development recognizes the inequitable distribution of environmental and economic burdens created by the development policies associated with industrial capitalism and colonial powers.

Additionally, the model of sustainable development would require governments to place constraints on development that have not been present before. Development is often a matter of constructing infrastructure such as roads, bridges and dams. Sustainable development requires the consideration of environmental and social concerns in such projects, not simply maximizing short-term economic benefits at the sacrifice of the environment or communities within that developmental area. Development also often means new manufacturing methods. Sustainable development requires commitment to the principles and practices of clean production and manufacturing techniques rather than continued reliance on fossil fuels, or other dirty energies to propel manufacturing.

Shifting models of development raise controversies that may involve problems of unequal opportunity for women and subordinated ethnic groups and the environmental impacts of industrial use of natural resources. These controversies stem from the development policies and practices of an earlier age that did not require accountability for the social or environmental consequences of development. The changed model has been hardest to accept in the United States.

CONTROVERSIES FOR BUSINESSES AND INDUSTRIES

Sustainable development requires businesses and industries to adopt clean manufacturing goals and technologies. This often means designing pollution and waste out of their manufacturing cycle (industrial ecology) and thinking about their product in terms of its total life span, beyond its point of sale (product life cycle management). Additionally, clean production may also require a substantial investment in new technology and plants, investment that is prohibitive to small business enterprises. These requirements challenge businesses and industries in virtually all sectors of an economy to change what they are doing and how they are doing it. What makes the task of change even harder is the fact that many established businesses and industries have been subsidized directly or indirectly through tax benefits conferred by national governments, which enhance the reluctance of businesses to change. But some businesses have pioneered change by embracing concepts of a restorative economy and natural capitalism.

Businesses that have taken short-term transitional losses to eliminate waste and toxins in their product and production methods have been rewarded in long-term economic gains, as well as creating measurable improvements in their environmental impacts. These businesses have embraced the linkage between environment and economy, but they have not necessarily incorporated communities and their well-being into this new model.

ENVIRONMENTALISTS

Environmentalists have documented the scope of environmental degradation all ecosystems are suffering as the result of human activities. The news they deliver is sobering. Human activity is threatening to cause the collapse of the living systems on which all life on our planet depends. This leads some environmentalists to advocate protection of the environment above all other concerns, including economic concerns and the needs of human communities. Trying to determine the causes for this dangerous state of the environment leads some environmentalists to identify population growth as the most substantial factor. Others point to use of fossil fuel to propel our activities. Still others identify overconsumption of resources as the basis for this state of environmental degradation.

Environmentalists employing such data and using this type of formula often find themselves in conflict with business and industry as they press urgently for changes in manufacturing processes. Mainstream environmentalists tend to be from relatively privileged backgrounds including wealth and educational opportunities. They frequently find themselves in conflict with communities and developing countries for their stands about population control and their relative indifference about the plight of the poor, who bear the costs of unsustainable practices more than any other class.

COMMUNITIES

Communities and their physical and economic well-being are often excluded from decision making concerning economic opportunities and environmental consequences with which they must live. This exclusion can arise from structural separation between different administrative branches of government, or from the

INDUSTRIAL ECOLOGY

Industrial ecology is the shifting of industrial process from open loop systems, in which resource and capital investments move through the system to become waste, to a closed loop system where wastes become inputs for new processes. Robert Frosch first proposed the idea of industrial ecology in an article published in *Scientific American* in 1989. He asked, "Why would not our industrial system behave like an ecosystem, where the wastes of a species may be resource to another species? Why would not the outputs of an industry be the inputs of another, thus reducing use of raw materials, pollution, and saving on waste treatment?" The idea of industrial ecology is to model this human-made system on the performance of those based on natural capital that do not have waste in them. The term industrial ecology was defined as a systematic analysis of industrial operations by including factors like technology, environment, natural resources, biomedical aspects, institutional and legal matters as well as socioeconomic aspects.

Industrial ecology conceptualizes industrial systems like a factory or industrial plant as a human-made ecosystem based on human investments of infrastructural capital rather than reliant on natural capital. Along with more general energy conservation and material conservation goals, and redefining commodity markets and product stewardship relations strictly as a service economy, industrial ecology is one of the four objectives of natural capitalism. This strategy discourages forms of amoral purchasing arising from ignorance of what goes on at a distance and implies a political economy that values natural capital highly and relies on more instructional capital to design and maintain each unique industrial ecology.

separation between different levels of government. It can also arise from cultural and social forces that formally or informally operate to exclude poor people or people stigmatized by historical discrimination. Exclusion of community interests and participation affects the viability and efficiency of efforts to protect the environment and to develop a community economically in several ways. People driven by insecurity as to basic living conditions are likely to accept employment opportunities regardless of consequences to human and environmental health. This eliminates labor as an agent of change toward sustainable production technologies and allows continued pollution and waste to be externalized into the environment with long-term disastrous consequences for human health. Moreover, people faced with exposure and hunger will also contribute to environmental degradation to meet basic life needs. Whether in a developed or a developing country, poverty and the inability to meet basic needs for food, shelter, and care make some human communities even more vulnerable to environmentally degraded conditions of work, living, recreation, and education.

Conflicts arise as these communities strive to participate in environmental decision making and decisions concerning the use of natural resources. These communities are often not welcomed into dialogue at a meaningful and early stage and are forced to seize opportunities to participate in controversial ways. In the United States, the environmental justice movement has pioneered processes of public participation designed to ensure community involvement with environmental decision making. Internationally, the United Nations has developed a convention to assure such participation, the Aarhus Convention. Banks and other international lenders are beginning to require such community participation in development projects. For example, the World Bank has now required community participation and accountability to communities in its lending programs under the Equator Principles. Additionally, people including ethnically stigmatized groups and women, disadvantaged by informal social forces are striving to use these methods, and others to participate in environmental and economic decision making. In the United States these efforts are being developed through the environmental justice movement. Internationally, the United Nations supports efforts to build strong nongovernmental organizations through which these and other community interests can be effectively championed. The UN activities are being developed through the Civil Society initiatives.

POTENTIAL FOR FUTURE CONTROVERSIES

There is a strong international and national push for a new kind of environmentalism that includes sustainable development. Like all the environmental policies before it, sustainable development policies will need to have accurate, timely, and continuous data of all environmental impacts to be truly effective. So far knowledge needs about environmental impacts generate strong political controversies. In the United States most of the industry information is self-reported, the environmental laws are weakly enforced, and environmental governmental agencies are new. Sustainability will be controversial because it will push open old controversies like right-to-know laws, corporate audit and antidisclosure laws, citizen monitoring of environmental decisions, the precautionary principle, true cost accounting, unequal enforcement of environmental laws, and cumulative impacts.

The concept of sustainability has captured the environmental imagination of a broad range of stakeholders. No one group is against it, in principle. It is the application of the principle that fires up underlying value differences and old and continuing controversies. In many ways, the strong growth of the principles of sustainability represents exasperation with older, incomplete environmental policies. These policies now seem piecemeal, ineffective in individual application, and an impediment to collaboration with other agencies or environmental stakeholders. The new processes of policies of sustainability could be radically different than environmental decision making is now. Communities are demanding sustainability, some even if adopting the Kyoto principles even if the United States will not sign it under President George W. Bush. They want to be an integral part of the process, especially as they learn about the land, air, and water around them. As environmental literacy spreads so too will all the unresolved environmental controversy. It is these controversies that lay the groundwork for the functional advancement of U.S. environmental policy. Sustainability will require complete inclusion of all environmental impacts, past, present, and future.

See also Citizen Monitoring of Environmental Decisions; Cumulative Emissions, Impacts, and Risks; Environmental Impact Statements: International; Precautionary Principle; True Cost Pricing in Environmental Economics

Web Resources

The Equator Principles. Available at www.equator-principles.com/index.html. Accessed January 22, 2008.

Union of Concerned Scientists. Available at http://www.ucsusa.org/. Accessed January 22, 2008.

The United Nations and Civil Society. Available at www.un.org/issues/civilsociety. Accessed January 22, 2008.

United Nations Economic Commission for Europe. The Aarhus Convention. Available at www.unece.org/env/pp/. Accessed January 22, 2008.

Further Reading: Apostolopoulos, Yiorgos, and Dennis John Gayle. 2002. *Island Tourism and Sustainable Development. Caribbean, Pacific and Mediterranean Experiences.* Westport, CT: Praeger; Dernbach, John C. 2000. *Stumbling toward Sustainability.* Washington, DC: Environmental Law Institute; Doob, Leonard William. 1995. *Sustainers and Sustainability: Attitudes, Attributes, and Actions for Survival.* Westport, CT: Praeger; Harris, Jonathan M. 2001. *A Survey of Sustainable Development: Social and Economic Dimensions.* MO: Island Press; Hawken, Paul. 1993. *Ecology of Commerce (The): A Declaration of Sustainability.* New York: HarperCollins; Maser, Chris. 1999. *Ecological Diversity in Sustainable Development: The Vital and Forgotten Dimension.* New York: Lewis Publishers; Rao, P. K. 2000. *Sustainable Development.* Oxford: Blackwell Publishing; Riddell, Robert. 2004. *Sustainable Urban Planning: Tipping the Balance.* Oxford: Blackwell Publishing; Wackernagel, Mathis, and William Rees. 1995. *Our Ecological Footprint: Reducing Human Impact on the Earth.* Gabriola Island, Canada: New Society Publishers.

Robin Morris Collin

T

"TAKINGS" OF PRIVATE PROPERTY UNDER THE U.S. CONSTITUTION

The role of government and its ability to take away property owned by private citizens is one of the foundations of the U.S. concept of liberty. The takings clause of the Fifth Amendment to the U.S. Constitution states: "nor shall private property be taken for public use, without just compensation." The government can take private property but it must be for a public purpose and they must pay you what they determine to be fair market value. This is called the *eminent domain* of the state to pursue the greater good. Environmentalists advocate for a stronger role for government to control environmentally harmful practices of private property owners. If governments can regulate land without taking away the property, then no compensation needs to be paid. Regulations that do essentially take the value of the property down to zero may require the government to pay compensation for lost value. Governments seek to avoid this but are under increasing pressure from community groups and other voters to stop environmentally nonsustainable and dangerous uses of private property. The crux of this deep institutional controversy is the need to act in an ecological manner in the face of political land control and distribution regimes.

Land-use law is a murky jurisprudence, with governments suing other governments, citizens seeking redress in the courts, and a strong focus on takings. Historically, it was unusual for the U.S. Supreme Court to even consider land-use cases, preferring to leave these to the state and local control. This U.S. Supreme Court has been more active.

Many state constitutions have dynamic takings provisions. Their interpretation has changed over time by different state courts. However, the U.S. Supreme Court has ruled that certain regulations can go too far and therefore result in takings that require the government to compensate the landowner if they need the land. In general, the Court has ruled that regulation only results in a taking when it eliminates all or substantially all of the property's value.

Business proponents of the modern takings agenda argue that regulations that limit the potential value of land and other property result in takings, which causes a loss of value that must be compensated for by the government. This includes many environmental rules and laws. These proponents do not address what happens when a government act increases the value of their land; they just assume it is their personal profit. An example of this is when the government builds a train station. All the retail trade property owners have a better site, and the value of the property goes up. The U.S. land-use system does not recognize givings as the English do.

The courts will play a powerful role in determining the direction of the taking issue because of its constitutional roots. This is true for several reasons:

- The U.S. Supreme Court is the final arbiter of the meaning of a constitutional provision such as the takings clause. Absent a further amendment to the Constitution (a rare event), citizens and their elected representatives are powerless to overturn the Court's interpretation of the Constitution.
- The courts are inherently conservative institutions. Once the Supreme Court establishes precedent in some field of law, the Court tends to revise its precedents, if at all, only reluctantly and over a relatively long period of time.

Over the last decade, the Supreme Court has taken a number of important steps in the direction of expanding the takings clause as a constraint on the ability of elected representatives to adopt regulations to protect the environment and other aspects of the public welfare.

OREGON'S TEMPORARY TAKINGS COMPENSATION

Measure 37

Oregonians approved Measure 37 in November 2004 by 61 percent to 39 percent. Oregon, like several other states and territories of the United States, votes on citizen initiatives if enough signatures from citizens are gathered. These initiatives often take the form of measures. In terms of policy, they vary widely in scope, content, and legality. Measure 37 allows a property owner to receive compensation for any land-use regulation passed since the owner or his family controlled the property or, alternatively, waive any land-use regulation enacted since the owner has had the property. All counties in Oregon have decided to use waivers as the primary way to address Measure 37 claims; some have reserved a small amount of money to compensate exceptional claims.

Oregon's Land-Use Planning System: An Early State Approach

In 1973, Senate Bill 100 was passed by the Oregon Legislature and signed into law, creating a statewide planning program that has since been expanded. Many statewide environmental groups, especially 1,000 Friends of Oregon, advocated for passage of this law. The planning program is coordinated by the state Land Conservation and Development Commission and the Department of Land Conservation and Development but implemented by local governments. Local governments develop growth and zoning plans that then must be approved by the state. The most controversial element of these plans is often the urban growth boundary, which limits how property can be sold or developed outside those boundaries. Many proponents of Measure 37 have objected to restrictions on building homes, including for owner-occupancy and subdividing.

As of 2006 all of Oregon's 240 cities and 36 counties have comprehensive land-use plans, every city area has an urban growth boundary that separates urban uses from rural farm and forest uses, and all sensitive coastal lands and estuaries on Oregon's coast are part of a comprehensive management plan. Using land-use planning, Oregon has concentrated population growth in cities: for instance, between 1979 and 1999, the Portland Metro area had a 40 percent increase in population and only a 20 percent increase in urbanized land, while between 1990 and 1996, Kansas City had a 5 percent increase in population and a 70 percent increase in urbanized land. The land-use plans have also maintained amounts of farmland at the same time many states, including Pennsylvania and Florida, lost significant amounts of farmland.

In addition to land-use planning regulations, Oregon also has a forestry law that will be involved in Measure 37 claims. The 1971 Forest Practices Act requires replanting of logged tracts, limits the size of clear-cuts, and places restrictions on activities near streams. This can be construed as a temporary taking of private property for which Measure 37 would mandate compensation to the property owner.

How Measure 37 Works

To enforce the act, owners are to write to the local or state government concerning the regulation they object to; the act calls for them to receive money or the regulation to be vacated within 180 days. Owners can only receive relief if they or a family member owned the property when the land restriction they object to was passed. If the government does not grant relief within 180 days, the statute creates a civil right of action and entitles the landowner to attorney's fees. The measure exempts from its compensation obligations, rules, and legislation that control common-law public nuisances, selling pornography, and nude dancing, or legislation that restricts activities to protect public health, protect safety, or comply with federal law.

Because Oregon's land-use planning program delegates development and implementation to local governments, they will be responsible for addressing most of these claims and paying for them. The state has estimated costs to local governments will be between $46 million and $300 million. The state has no obligation to cover those costs. Local governments can lower their costs by relaxing restrictions instead of paying compensation.

More than 1,000 Measure 37 claims have been filed since December. Many are requests to build an additional home on an area zoned for farming. Some of the larger claims are detailed here.

Claim against Protections around Native American Sacred Places

Wallowa County: A Measure 37 claim has been filed by a developer who objects to density restrictions on his subdivision. The density restrictions had been imposed because of the Nez Perces' objection to the subdivision, which they found to be too close to the grave and monument of Old Chief Joseph of the Nez Perce tribe, culturally significant land and the site of many possible archeological sites. The developer had been permitted to build 22 homes on 60 acres but now wants to build 60–72 homes on the same land.

Claim to Build Casino in a Town of 400

Marion County: Two brothers have filed for permission to build a casino, hotel, and golf course on their farm outside a town of 400. Fire Chief Reed Godfrey of the St. Paul fire district that would be required to service the McKay casino development, has said that the project "would completely overwhelm our fire department and ambulance transport capabilities." While the Oregon Constitution prohibits nontribal casinos, the brothers claim their family has owned the land since before Oregon became a state.

Claim to Build Project Rejected by Community as Too Noisy

Lane County: A Measure 37 claim to create 300-acre motorcycle, ATV, and paintball park that has already been rejected by the local county council was filed in late January. Neighbors have been fighting the proposal for years, and the county council had turned it down because of noise pollution and fire hazard concerns.

Claims for Subdivisions in the Columbia River Gorge

Columbia River Gorge is a federally designated scenic area. Land-management decisions are federally designated to the interstate Columbia River Gorge Commission; that commission has delegated some of its authority to local counties. Despite language in Measure 37 excluding regulations designed to comply with federal law, several Measure 37 claims now argue that the gorge regulations are state and county regulations and can be waived. Because the Columbia Gorge building restrictions are fairly recent (1986), many landowners would qualify for waivers and thus be allowed to turn their property into subdivisions. One claim to build hundreds of homes asks for $15.6 million. In April, the Columbia River Gorge asked a state court whether Measure 37 applies to their regulations.

Claim for a Right to a Landfill

Washington County: A 43-acre landfill filed a Measure 37 claim. The company asks that the state waive its 209-foot height limitation, as well as allow the landfill to extend northwards, or pay the company $11.4 million. The landfill is already unpopular with neighbors.

Claim to Build a Mall in the Woods

A Measure 37 claim in rural Polk County has been approved for over 1 million square feet of commercial space, approximately the size of Oregon's largest malls.

Subdivision Development Claims

- 842 homes on one-quarter-acre lots or receive $57 million in the Hood River valley
- 340 acres of housing and commercial development in Yamhill County (April 5, 2005)
- 167 acres of housing on 10-acre lots in Yamhill County (April 16, 2005)
- 173 acres of farmland (in heard) of Red Hills wine country Yamhill County; alternately are seeking $15.6 million (January 27, 2005)

Pioneering Controversies

In Measure 37, counties are given the choice between paying compensation for the loss of property value because of restrictions on the property since a family has held it or granting waivers of the law since the claimant acquired the property. Most counties in Oregon have stated that they will grant waivers because paying compensation is simply impossible; in Yamhill County two of 74 claims were for more than the county's entire yearly property tax revenue. Those whose family has owned the property for a long time but have only recently inherited it themselves are left in the middle. In Linn County, a family in such a position is pressing the county to waive laws enacted before they inherited the property. On May 25, 2005, the county council turned down the family's waiver request but the family is expected to litigate.

The exception in Measure 37 for "the protection of public health and safety" is likely to be controversial. Some issues that have already been raised are whether setbacks from forested lands can be waived or if they are a health and safety measure designed to limit forest fires, and whether limitations on building because of limited groundwater can be justified as a health restriction.

Whether Measure 37 requires continual ownership is also debated. Claims have been rejected because there was not continual ownership, but in Linn County a widow whose husband had owned the property for many years legally by himself was allowed to receive a waiver from regulations enacted when her husband owned the property.

Other potential controversies will arise as Measure 37 is implemented. The measure is expected to impact logging practices, but some of these could be justified under federal statutes like the Endangered Species Act. Many counties have decided to waive rules for all valid Measure 37 claims and not even consider compensation. If Measure 37 claims are filed where only a few hundred dollars or several thousand dollars are lost in value and the county waives important land-use regulations, more controversies could be created. If compensation is to be paid, no one knows quite how it should be determined.

Measure 37 and the case law it will spawn are being closely watched by many private property rights activists, developers and land speculators, environmentalists, state and local governments, and land-use planners. It is likely that some aspects of Measure 37 will attempt to reach the U.S. Supreme Court. It evokes a controversy about a core American value held from the homeowner level and determined by the highest court—private property.

The activity at the Supreme Court level is matched by a significant rise in the volume of takings-related litigation in the lower courts. The U.S. Court of Appeals for the Federal Circuit and the U.S. Court of Federal Claims, in particular, have been especially active in recent years in attempting to expand regulatory takings doctrine. The state as well as federal courts has seen a significant increase in the number of takings cases filed in recent years. In Congress, takings legislation was a central feature of the Contract with America in the 104th Congress. Some version of takings legislation has been considered, but not adopted, in every subsequent Congress. Virtually every state legislature has considered takings legislation; about 20 states have adopted some form of legislation on this subject. In the state legislative sessions of early 2005, a number of takings bills have been introduced. Some are in response to Measure 37's passage in Oregon; others are perennially introduced, as in Maine. Ballot initiatives like Measure 37 are also being discussed in a number of states.

The takings clause of the Fifth Amendment is one of the few provisions of the Bill of Rights that is very important to free-market conservatives. Government regulation is a concern of theirs generally, but private property regulation is an issue of strong political saliency.

It is clear that when the government physically seizes property (as for a highway or a park, for example) that it will have to pay just compensation. It is also clear that serious, sustained physical invasions of property (as in the case of low overflying aircraft, for example) require payment of compensation equal to the difference between the market value before and after the invasion. The difficult cases are generally those where government regulations, enacted to secure some sort of public benefit, fall disproportionately on some property owners and cause significant diminution of property value. The Court has a difficult time articulating a test to determine when a regulation becomes a taking. It has said there is "no set formula" and that courts "must look to the particular circumstances of the case." The Court has identified some relevant factors to consider: the economic impact of the regulation, the degree to which the regulation interferes with investor-backed expectations, and the character of the government action. There is a lot of room for controversy as to how these various factors should be weighed.

The takings law is in the courts and legislatures. The battles are expensive, intense, and increasing. So too are the environmental controversies that drive public concern over ecological assessments, impacts, and risks. Sustainability may require a significant shift in the direction of takings law and policy. If one private property owner is free to pollute an entire ecosystem, then current takings law may not be sustainable. Large changes in takings law loom in light of global conditions and cumulative industrial impacts. These changes will upset the settled expectations investors and owners have regarding real property. The next set of controversies will be the valuation of theses types of partial property losses.

POTENTIAL FOR FUTURE CONTROVERSY

While courts may engage this controversy for many years and issue decisions, and legislatures may deliberate and hold hearings, and administrative agencies may take years to promulgate very important environmental rules, the impact on the environment continues. The fundamental association of private property and liberty is one that has shifted dynamically over the history of the United States. English systems recognize not only takings but also givings, where the property owner is charged for the increased value of the land due to government services. As increasing environmental impacts and knowledge about them continue unabated, the strong U.S. association of liberty with private property will be challenged. This controversy will challenge the foundations and flexibility of concepts of liberty and private property.

See also Federal Environmental Land Use; Land-Use Planning in the United States; State Environmental Land Use; Sustainability

Web Resources

The Property Rights Project. Available at www.propertyrightsproject.org/. Accessed January 22, 2008.

U.S. Environmental Protection Agency. Wetlands: What about Takings? Available at www.epa.gov/owow/wetlands/facts/fact18.html. Accessed January 22, 2008.

Further Reading: Bradley, Jennifer, Timothy Dowling, and Douglas Kendall. 2006. *Good News about Takings.* Chicago: APA Planners Press; Epstein, Richard. 1985. *Takings: Private Property and the Power of Eminent Domain.* Cambridge, MA: Harvard University Press; Freyfogle, Eric. 2003. *The Land We Share: Private Property and the Common Good.* Washington, DC: Island Press; Hagman, Donald G., Dean J. Misczynski, and U.S. Department of Housing and Urban Development, Office of Policy Development and Research. 1978. *Windfalls for Wipeouts: Land Value Capture and Compensation.* Chicago: American Planning Association; Pruetz, Rick. 2003. *Beyond Takings and Giving.* Marina Del Ray, CA: Arje Press.

TOTAL MAXIMUM DAILY LOADS (TMDL) OF CHEMICALS IN WATER

The overall controversy of setting standards for water quality has not been resolved in the United States. There are some basic standards for a small number of chemicals discharged into the water. Total maximum daily loads (TMDLs) represent the best efforts to set comprehensive water quality standards. It is a calculation of the maximum amount of a pollutant that a water body can receive and still meet water quality standards. It determines the allocation of that amount to the pollutant's sources among all water permit holders. Water quality standards are legally set by states, territories, and tribes. These standards identify the uses for each water body, for example, drinking water supply, contact recreation (swimming), and aquatic life support (fishing). Most sources of water pollution are from nonpoint sources so even if TMDLs are established it will be hard to assess responsibility for discharges into water. Environmentalists and communities want water standards that disclose the total maximum load of a given chemical, and how much capacity the watershed has for that chemical. TMDL standards have been conceptual environmental policy tools for decades but have faltered at the implementation stage. Industry strongly resists the formulation of such standards because it increases the cost of pollution controls.

The environmental law mandate to provide safe, drinkable, fishable, and swimmable water flounders on trying to control environmental impact on a pollutant-by-pollutant or chemical-by-chemical approach. Theoretically, a TMDL is the total of the allowable loads of a single pollutant from all contributing point and nonpoint sources. Nonpoint sources is a catchall term for sources that are currently not known or measured. The calculation should include a margin of safety. The calculation also accounts for seasonal variation in water quality. Both these calculations can be controversial depending on stakeholder perspective. The Clean Water Act, section 303, establishes the water quality standards and TMDL programs. The standards can be clearly described by the federal government, but states are free to try new ways to achieve these standards. Calculations of margin of safety and seasonal variations are common battlegrounds in this controversy, partially because of this dynamic of intergovernmental relations in U.S. environmental policy. Unlike the Clean Air Act that divided the country into about 290 districts and measured six basic pollutants, U.S. clean water policy is still at the stage of describing how much of a given chemical or potential pollutant a given body of water can absorb. Some chemicals have definite effluent and discharge limitations per facility, but many do not. This battleground around standards of capacity for water bodies has continued since the early 1980s and is not close to resolution.

THE NEED: THE QUALITY OF OUR NATION'S WATERS

Almost one-half of assessed waters still do not meet the water quality standards that states, territories, and authorized tribes have set for them. This is over 20,000 individual river segments, lakes, and estuaries. These impaired waters include about 300,000 miles of rivers and shorelines and approximately five million acres of lakes. About 218 million people live within 10 miles of the polluted water.

Many aspects of the watershed, including other lakes, rivers, and streams, have not been environmentally assessed. There are still many controversies about the underlying standards for water quality measurement and measurement protocol. Some states allow chemical discharges into water to mix with river water before they are tested. Other states have claimed water quality testing protocols where none existed. The level of distrust is very high. This can prompt citizen monitoring of environmental conditions.

FEDERAL AGENCY BATTLEGROUND

Communities, states, and environmental groups blame the EPA for delays in issuing TMDL guidance and providing assistance to the states. The EPA and others are critical of states for vigorously exploring approaches to address their water quality problems unique to their states. A current political controversy involves progress in implementing the nonpoint pollution-management provisions added in 1987. States are developing management programs describing methods that will be used to reduce nonpoint pollution, which may be responsible for as much as 50 percent of the nation's remaining water quality problems. The EPA has adopted program guidance intended to give states more flexibility and to speed up progress in nonpoint source control. Nonpoint sources are those sources not covered by the Clean Water Act. Another political controversy is impacts and implementation of requirements under current law for states to develop TMDLs to restore pollution-impaired waters.

TRIBES AND TMDLs

Many Native American tribes have status as states with the U.S. Environmental Protection Agency. This is generally in accordance with long-standing treaty rights and other binding and public agreements made with tribes. If tribes meet certain conditions they can set their own TMDLs. This affects the ability of other water users in controversial ways. Of the approximately 500 tribes, about 10 percent either have or are seeking status as states. Many tribes have sophisticated science officers and a deep ethic of preservation of the natural landscape. Water quality is a deep and important multidimensional issue for indigenous people everywhere, affecting relationships with religion, survival, nature, and community. Many tribes also have manufacturing entities that may emit into the water. Some feel that tribes could set the standards so that other water users suffer. While water rights cases can be very complex, many cases involving tribal water rights mask other controversies that relate to unregulated growth without knowledge or accommodation of limited water resources. When resources get low all stakeholders exert their strongest legal claims. Some political claims by farmers, agribusiness, ranchers, loggers, and miners may not have the force of law, but they have had that effect. In this controversy, these cases continue right up to the U.S. Supreme Court.

A 1998 Supreme Court decision reaffirms a 2,500-member tribe's right to tell the city of Albuquerque what it can and cannot dump into the Rio Grande River. The Isleta Pueblo is located six miles downstream from where Albuquerque dumps 55 million gallons of wastewater each day. Sewage from the city's 450,000 residents makes the river water unhealthy for farming and religious ceremonies. Of particular concern are ammonia, a by-product of human waste, and arsenic, which comes out of the ground in city wells. Arsenic magnifies itself as it works its way up the food chain, poisoning fish and Isleta's centuries-old fields of squash and corn. The tribe had little legal clout with Albuquerque until 1987, when Congress amended the Clean Water Act, granting 129 tribes around the country equal standing with states on water quality issues. Native Americans had the right to dictate upstream water quality in rivers that flow through tribal lands.

Isleta was the first tribe to establish a water quality standards program as allowed under the act. It was the only opportunity to do something about the pollution legally. Isleta set a strict arsenic limit of 17 parts per trillion, many times cleaner than the federal drinking water standard of 50 parts per billion. Some observers say that setting the standards extra high was intentional, a bargaining wedge to negotiate a happy medium with the city.

At first, Albuquerque seemed cooperative so Isleta gave the city a three-year grace period to analyze its discharges into the Rio Grande and upgrade its sewage treatment. But in 1992, after the agreement was reached, the city sued the federal Environmental Protection Agency. Albuquerque argued it should not have to pay to clean up pollution when some of it came from industries polluting upstream or from naturally occurring arsenic.

Over five years, Albuquerque tried to convince two federal courts that Isleta's request had no scientific basis and that a clause in the Clean Water Act directs the EPA to mediate disputes over water quality standards. In October 1996, the 10th Federal Court of Appeals in Denver ruled against the city. It held the tribe had legal authority to enforce water quality. The city appealed the case to the Supreme Court, and when the court declined to hear the case last November 2006, it upheld Isleta's right to dictate Albuquerque's water quality.

The city will have to spend more than $300 million to upgrade its sewage plant. The plant could cost $20 million a year to operate. Currently, tribal water quality standards are being challenged in two states—in Wisconsin for the Mole Lake Chippewa and in Montana for the Confederated Salish and Kootenai Tribes and Fort Peck Assiniboine and Sioux tribes. In New Mexico, seven other tribes, many of which sit along major rivers, have developed water quality standard programs.

CURRENT STATUS

The EPA reported in the 2000 National Water Quality Inventory Report that 39 percent of assessed river and stream miles and 45 percent of assessed lake acres do not meet applicable water quality standards and were found to be impaired for one or more desired uses. The types of remaining water quality problems are diverse, ranging from runoff from farms and ranches, city streets, and other diffuse sources, to metals (especially mercury), organic and inorganic toxic substances discharged from factories, sewage treatment plants, and non-point sources.

The Bush administration has been reviewing a number of current clean water programs and rules but has proposed few new initiatives regarding the implementation of TMDLs.

POTENTIAL FOR FUTURE CONTROVERSY

TMDL implementation will involve individual landowners and public or private industries in agriculture, forestry, and urban development. They become very concerned about TMDLs because it affects the cost and availability of water. Some are concerned with the government destroying the value of their land if they are not allocated a certain amount of water and view that as a taking of private property. In these cases, the controversy will move into judicial battlegrounds.

Meanwhile, the TMDL policy flutters between state and federal environmental agencies, moving slowly. More community groups are concerned about the lack of any applicable standards for maximum daily loads, or loads of any type, for chemicals in their water. The controversy about TMDLs is not about a way to set standards for water quality but about setting them at all. Resistance to setting standards prevents measuring changes in the environment. This makes it difficult to know what will work, which slows policy development and pulls in other scientific controversies. Meanwhile, the water quality suffers from many users.

See also Citizen Monitoring of Environmental Decisions; Permitting Industrial Emissions: Water; Public Involvement and Participation in Environmental Decisions; Water Pollution; Watershed Protection and Soil Conservation

Web Resources

Association of Environmental Authorities of New Jersey. Recent TMDL decision under Consideration for Supreme Court Review. Available at www.aeanj.org/Main%20 EPA%20News.htm. Accessed January 22, 2008.

The Future of the TMDL Program: How to Make TMDLs Effective Tools for Improving Water Quality. House of Representatives, 107 Cong., 1st Sess., November 15, 2001. Available at commdocs.house.gov/committees/trans/hpw107–56.000/hpw107–56_0.HTM. Accessed January 22, 2008.

Further Reading: Commission on Geosciences, Environment and Resources. 2001. *Assessing the TMDL Approach to Water Quality Management.* Washington, DC: National Academy Press; Houck, Oliver A. 1999. *The Clean Water Act TMDL Program: Law, Policy, and Implementation.* Washington, DC: Environmental Law Institute; Younos, Tamim, ed. 2005. *Total Maximum Daily Load: Approaches and Challenges.* Tulsa, OK: PennWell Books.

TOXIC WASTE AND RACE

Environmental justice includes the distribution of environmental benefits and burdens. One of the most unwanted burdens is toxic waste. The race of the community is one of the best predictors of where controlled and uncontrolled hazardous wastes exist.

The United States rose quickly to large-scale industrialization. Its population grew as fast as industry. There was little land-use control until 1915. Industry, business, and technology were all highly encouraged and seen as an indicator of healthy prosperity. Many cities became the focal point for waves of immigration and post–Civil War migration. Conditions were often overcrowded and unsanitary. Diseases were frequent and more deadly. Racism in housing, employment, voting, and education was legal during most of this period. With former slaves facing the fury of Reconstruction many moved to northeastern cities. After World War II many whites left the city for the suburbs. Postwar economic robustness fueled a rapid increase in automobile and truck transportation. The waste from a century of industrialization was often left in place, in the city. The controversy about toxic waste increased as knowledge increased about the potential and real risks posed by it. Our society shows no signs of decreasing the production of toxic wastes. Our population is increasing, putting waste and communities in direct conflict. And our knowledge base about the public health consequences of toxic waste continues to grow as citizens become active. All these growing and dynamic factors contribute to controversy.

In the 1970s and later, federal and state governmental agencies were formed to regulate all kinds of waste and hazardous materials. As these agencies studied the distribution of toxic waste facilities they found a pattern. At the request of District of Columbia Congressman Walter Fauntroy, the U.S. General Accounting Office conducted a study of eight southern states (Alabama, Florida, Georgia, Kentucky, Mississippi, North Carolina, South Carolina, and Tennessee) to determine the correlation between the location of hazardous waste landfills and the racial and economic status of the surrounding communities. The results showed a clear bias in landfill placement, with three out of every four landfills sited near predominantly minority communities. The U.S. Environmental Protection Agency (EPA) was not formed until 1970. Many state environmental agencies and most regional planning agencies were created since then. The Office of Environmental Justice was created inside the EPA in 1991. Most States now have some agency or task force on environmental justice.

WHAT IS TOXIC?

By law, the Agency for Toxic Substances and Disease Registry (ATSDR) produces toxicological profiles for hazardous substances found at National Priorities List (NPL) sites. These hazardous substances are ranked based on frequency of occurrence at NPL sites, toxicity, and potential for human exposure. Toxicological profiles are developed from a priority list of 275 substances. Also by law, ATSDR prepares toxicological profiles for the Department of Defense (DOD) and the Department of Energy (DOE) on substances related to federal sites. According to the ATSDR, 289 toxicological profiles have been published or are under development as finals or drafts for public comment; 268 profiles were published as finals; 118 profiles have been updated. Currently, 14 profiles are being revised based on public comments received and 7 profiles are being developed as drafts for public comment. These profiles cover more than 250 substances, but there are many more substances to research. Chemical manufacturers resist the label of toxic or hazardous in agency and legislative hearings. The amount of scientific uncertainty looms large now but is decreasing as information about the environmental impacts of contested toxics becomes known. Controversies ensue because citizens are uncomfortable about being asked to assume the risk of this scientific uncertainty and would rather wait until a given substance is proven safe.

The law now requires ATSDR to provide toxicological profiles to state health and environmental agencies and to make them available to other interested parties.

Toxics are also listed on the Toxics Release Inventory (TRI), again by law. Begun in 1988, the Toxics Release Inventory contains information on releases of nearly 650 chemicals and chemical categories from industries including manufacturing, metal and coal mining, electric utilities, and commercial hazardous waste treatment, among others. As this information became known to academic and government researchers, more conclusions were drawn about toxic waste and race. As the information became readily available to citizens, groups exposed to waste and pollution confronted some of these environmental policies.

According to U.S. federal environmental law, hazardous waste is waste with properties that make it dangerous or potentially harmful to human health or the environment. The universe of hazardous wastes is large and diverse. Hazardous wastes can be liquids, solids, contained gases, or sludges. They can be the by-products of manufacturing processes or simply discarded commercial products, like cleaning fluids or pesticides. In regulatory terms, a hazardous waste is a waste that appears on one of the four lists of hazardous wastes or exhibits at least one of four characteristics—ignitability, corrosivity, reactivity, or toxicity. By definition, the EPA determined that some specific wastes are hazardous. These wastes are incorporated into lists published by the EPA. These lists are organized into three categories by EPA rule.

1. The F-list (nonspecific source wastes). This list identifies wastes from common manufacturing and industrial processes, such as solvents that have been used in cleaning or degreasing operations. Because the processes producing these wastes can occur in different sectors of industry, the F-listed wastes are known as wastes from nonspecific sources.
2. The K-list (source-specific wastes). This list includes certain wastes from specific industries, such as petroleum refining or pesticide manufacturing. Certain sludges and wastewaters from treatment and production processes in these industries are examples of source-specific wastes.
3. The P-list and the U-list (discarded commercial chemical products). These lists include specific commercial chemical products in an unused form. Some pesticides and some pharmaceutical products become hazardous waste when discarded.

Ignitable wastes can create fires under certain conditions, are spontaneously combustible, or have a flash point below 60°C (140°F). Examples include waste oils and used solvents. Corrosive wastes are acids or bases (pH less than or equal to 2, or greater than or equal to 12.5) that are capable of corroding metal containers, such as storage tanks, drums, and barrels. Battery acid is a common example. Reactive wastes are unstable under normal conditions. They can cause explosions, toxic fumes, gases, or vapors when heated, compressed, or mixed with water. Examples include lithium-sulfur batteries and explosives. Toxic wastes are harmful or fatal when ingested or absorbed (e.g., containing mercury, lead, etc.). When toxic wastes are disposed of in land, contaminated liquid may leach from the waste and pollute groundwater.

There is little question about the presence and expansion of legal and illegal hazardous waste sites. These contain lead, mercury, arsenic, petroleum by-products, solvents, silicon, and many other known toxic substances. Grass clippings, disposable diapers, and other household waste that enter a legal nonhazardous waste site may not yet pose a toxic threat, and these are not included in the analysis that follows. It is likely that these waste disposal practices follow the same patterns as legal and illegal hazardous waste.

The United Church of Christ's Commission on Racial Justice became very involved in the presence of waste sites in communities of color. In 1987, the commission published a powerful study, "Toxic Waste and Race in the United States." This report demonstrated that race was the most significant factor in determining the siting of hazardous waste facilities, and that three out of every five African Americans and Hispanics live in a community housing toxic waste sites. The commission also noted that African Americans were heavily over-represented in areas neighboring toxic waste sites. A follow-up study in 1994 found the risks to be even greater: People of color are 47 percent more likely than whites to live near health-threatening facilities. In the early 1990s the *National Law Journal* reported that Superfund toxic waste sites in communities of color are likely to be cleaned up by the government 12 to 42 percent later than sites in white communities. This first round of research, study, and policy review fueled more controversy.

WASTE AND RACE

Efforts by citizens and government agencies to address environmental injustices have been hampered by a lack of public attention and research on the issue. New methods are being developed that rank the environmental burden of every community. They also measure cumulative exposure to environmental hazards of all kinds. Most other environmental justice studies focus on a particular type of hazard or facility. The findings demonstrate that environmentally hazardous facilities and sites—such as toxic waste dumps, polluting industrial plants, incinerators, power plants, and landfills—are disproportionately located in communities of color and low-income neighborhoods. Statistics present a stark imbalance. Communities of color average 27 hazardous waste sites per square mile (psm), while low-income communities average 14 waste sites psm. In contrast, middle-to-upper-income white communities average only 3 sites psm.

Similarly, between 1990 and 1998, large industrial facilities released an average of 110,000 to 123,770 pounds of chemical pollutants psm into communities of color, compared with 22,735 pounds psm for predominantly white communities. Such disparities are the result of political disempowerment and economic abandonment.

The landmark study, "Toxic Wastes and Race in the United States" (Commission for Racial Justice, United Church of Christ, 1987), described the extent of environmental racism and the consequences for those who are victims of polluted environments. The study revealed that:

- Race was the most significant variable associated with the location of hazardous waste sites.
- The greatest number of commercial hazardous facilities were located in communities with the highest composition of racial and ethnic minorities.

- The average minority population in communities with one commercial hazardous waste facility was twice the average minority percentage in communities without such facilities.

Although socioeconomic status was also an important variable in the location of these sites, race was the most significant even after controlling for urban and regional differences. The report indicated that three out of every five black and Hispanic Americans lived in communities with one or more toxic waste sites. Over 15 million African American, over eight million Hispanics, and about 50 percent of Asian/Pacific Islanders and Native Americans are living in communities with one or more abandoned or uncontrolled toxic waste sites.

A study by the EPA concluded that socioeconomic conditions and race are the major factors determining environmental discrimination. Communities inhabited by poor whites are also vulnerable to toxic threats. In its two-volume report, "Environmental Equity" (1992), the EPA alluded to the difficulties of assessing the impact of environmental hazards on low-income and minority communities. While admitting that those communities suffer a disproportionate share of the burden, there appears to be a general lack of data on the health effects of pollutants in those communities. The report asserts that environmental and health data are not routinely collected and analyzed by categories of income and race. Critics maintain that the information is available, but the EPA considers it a public relations issue, not a civil rights issue, and, therefore, does not take the claims seriously enough to gather the necessary data by income and race. This is a continuing battleground at the EPA and state environmental agencies.

One of the sources of toxic waste exposure is through incinerators. The ash that remains and the ash that flies into the air can contain metals and other chemicals that can pose a risk to people. The waste that is brought into the incinerator can also contain hazardous chemicals. This source of toxic waste is predominantly in racial minority communities in the United States. The portion of minorities living in communities with existing incinerators is at least 75 percent higher than the national average. In the late 1980s and early 1990s, communities where incinerators were proposed had minority populations 60 percent higher than the national average and property values 35 percent lower than the national average. This did not include Native American reservations where the majority of new incinerator proposals developed. In communities with existing incinerators, the average income is about 15 percent less than the national average and property values are 35 percent lower than the national average.

COMMUNITY CONCERN FOR THE ENVIRONMENT

Over 100 research and community studies find relationships between a high concentration of minority populations and pollution exposures from the environment. Industries that pollute are attracted to poor neighborhoods because land values, incomes, and other costs of doing business are lower. Higher-income areas are usually more successful in preventing or controlling the entry of polluting industries to their communities. The effects of pollution and environmental hazards on people of color have been overlooked because it was perceived that those communities would not successfully resist the siting of such facilities.

POTENTIAL FOR FUTURE CONTROVERSY

As our environmental policy evolves to include cities, the waste and race controversies will also include more cleanup controversies. The racism in the United States, from slavery until the present, is a value most now find repugnant. Implicit in any charge of racism is the idea that one must intend it to be racist. Historically, it may have been easier to prove racism in this manner than today, but it does not matter to any particular ecosystem. If the material there is toxic, it should be cleaned up. Toxic materials can accumulate over the decades and begin to move around in water and soil systems. The typical landfill liner is functional for only 30 years before the toxic leachate begins to seep into the soil and water. Who pays for the costs of a given cleanup is very controversial. The science and technology of waste disposal is also fraught with uncertainties. Who should pay for the cost of research and development in these areas?

These materials are toxic because of the damage they cause to living things. This can include cancers as well as a host of other debilitating effects. The damage to people is severe. Who pays for their costs? It is often difficult in our courts to prove, by a preponderance of the evidence, that any one specific act caused another individual to get cancer. Sometimes the period of time to sue, known as the statute of limitations, runs out before any affliction is reported. This is another controversial aspect of toxic waste and race—the damage to people and the costs to society.

Policy makers will have to specifically consider race when making environmental decisions about waste disposal. As U.S. environmental policy enters the multicultural urban arena, the cumulative totals of waste exposure will become more apparent. The City of Jackson, Florida, recently settled a $200 million lawsuit for $75 million where 4,500 people, predominantly African American, had been exposed to seven old-fashioned incinerators for decades. There was much harm to individuals. Individual health risks, reduced local property values, noxious fumes, increased truck traffic, and sometimes the attraction of hazardous waste–generating industry are also some of the burdens associated with these sites. Many cities may face similar histories and will face the controversies generated by waste and its legacy of placement in communities of color. The battleground for this controversy is expanding beyond environmentally disadvantaged communities. Concern about sustainability and ecosystems combined with spreading cumulative impacts on the public and on the environment focus a great deal of attention on toxic waste.

See also Cumulative Emissions, Impacts, and Risks; Ecosystem Risk Assessment; Environmental Justice; Human Health Risk Assessment; Toxics Release Inventory

Web Resources

Environmental Justice Resource Center. Available at www.ejrc.cau.edu/. Accessed January 22, 2008.

Toxic Waste and Race: Dangerous Blend. Available at www.finalcall.com/perspectives/env_racism05–07–2002.htm. Accessed January 22, 2008.

Further Reading: Bullard Robert D., ed. 2005. *The Quest for Environmental Justice: Human Rights and the Politics of Pollution.* San Francisco: Sierra Club Books; Goldman, Benjamin A., and Laura Fitton. 1994. *Toxic Wastes and Race Revisited: An Update of the 1987 Report of the Socioeconomic Characteristics of Communities with Hazardous Waste Sites.* Washington, DC: Center for Policy Alternatives; United States General Accounting Office. 1983. *Siting of Hazardous Waste Landfills and Their Correlation with Racial and Economic Status of Surrounding Communities.* June 1, 1983. GAO-RCED-83–168, B-211461.

TOXICS RELEASE INVENTORY

The Toxics Release Inventory (TRI) is a publicly available EPA database that contains information on toxic chemical releases and other waste management activities reported annually by certain covered industry groups as well as federal facilities. This is a powerful information tool in environmental controversies. There are current controversial proposals to change reporting requirements to every two years. There are always controversies about lack of coverage of all emitters of pollution and other chemicals.

An early stage of environmental policy formulation is to simply inventory or list and describe the places pollution comes from. *Pollution* is a term of policy art and has a much more limited meaning than most citizens expect. The chemicals and their amounts may qualify as pollution, but they may also be legally permitted by governmental environmental agencies. This definitional disjuncture is often the beginning of a battleground between citizens and communities, especially environmental justice communities, with state and federal environmental regulatory agencies like the U.S. Environmental Protection Agency (EPA). In part because of the rigid and entrenched battle lines with industry, elected officials passed laws that inventory the chemicals and wastes.

Another background factor with the formation of the TRI is that it came as the first really large cleanup policies began. Superfund and other large industrial cleanups began to untangle exactly who was responsible for the pollution. This is extremely difficult in some environmentally degraded sites with years of multiple industries, and sometimes municipalities, dumping large quantities of unknown chemicals into a site. This difficulty also helped develop a policy response of just creating an inventory of the sources of the chemicals and pollution. The inventory is a bitterly fought controversy because it is so powerful. Citizens can find out the self-reported emissions of some of the largest emitters.

The TRI was established under the Emergency Planning and Community Right-to-Know Act of 1986 (EPCRA) and expanded by the Pollution Prevention Act of 1990. EPCRA's primary purpose is to inform communities, first responders, and citizens of chemical hazards in their areas. Sections 311 and 312 of EPCRA require businesses to report the locations and quantities of chemicals stored on-site to state and local governments in order to help communities prepare to respond to chemical spills and similar emergencies. EPCRA Section 313 requires the EPA and the states to annually collect data on releases and transfers of certain toxic chemicals from industrial facilities and to make the data available to the public in the TRI. In 1990 Congress passed the Pollution Prevention Act that required additional data on waste management and source reduction activities to be reported under TRI. The express legislative goal of TRI is to empower citizens, through information, to hold companies and governments accountable in terms of how toxic chemicals are managed. Even though most of the TRI data is self-reported by covered industries, it is the best most groups can get or afford. It is heavily used. The more it is used, the more sources of emissions citizens want covered. The reason many citizens and communities want more industries included is to know more about their environment and any potential risks to themselves or the environment. Citizens, environmentalists, and others strongly believe that all environmental transactions should be transparent, that is, nothing with any potential environmental impact in any way is hidden. Sustainability proponents need all environmental information about a given ecosystem to better repair and manage it. Industries and local, state, and federal government agencies that protect them and hide their information resist disclosing all information even if it is self-reported. They would like to keep environmental audits secret and have succeeded in doing so in some states with the audit privilege.

The TRI program has expanded since its inception in 1987 to approximately 650 chemicals. Seven new industry sectors have been added to expand coverage significantly beyond the originally covered industries, that is, manufacturing industries. Most recently, the EPA has reduced the reporting thresholds for certain persistent, bioaccumulative, and toxic (PBT) chemicals in order to be able to provide additional information to the public on these chemicals. TRI listings are strongly resisted by covered and currently uncovered industries. Over four billion pounds of toxic chemicals are released into the U.S. environment each year, including 72 million pounds of recognized carcinogens, from nearly 24,000 industrial facilities. Environmentalists and many communities claim it is actually a much higher number in terms of real environmental impact. The information is self-reported for the most part. It is in industry's profit interest to view its emissions as too low to even cross the threshold necessary to begin reporting. It is a matter of debate as to whether some industries expand their plant by developing many small operations to keep from having to report anything. For example, in the late 1990s when New York City began requiring reporting of incinerator emissions only for those a certain size and over, many small incinerators sprung up with no reporting. Incinerator emissions are often toxic and contain metals that float down nearby as particulate matter. New York City is one of the most densely populated areas in the United States. The environmental impact could be even greater and more risky for people unless all environmental emissions are included in the inventory, argue environmentalists. They also argue that it is just an inventory, which by itself does not reduce environmental impacts. This is a constant battleground, driven by cumulative emissions and expanding public environmental literacy and concern.

A CONTROVERSIAL CHANGE IN REPORTING

In the October 4, 2005, Federal Register, the EPA noted that it wanted to make changes to the Toxic Release Inventory Reporting Program. The EPA has proposed three changes in TRI reporting.

1. Move from the current annual reporting to every-other-year reporting for all facilities, essentially eliminating half of the TRI program.
2. Allow companies to release 10 times as much pollution (raising the reporting threshold from 500 pounds to 5,000 pounds) before requiring them to report on how much pollution was produced and where it went.
3. Allow facilities to withhold information on low-level production of persistent bioaccumulative toxins (PBTs), including lead and mercury, which are dangerous even in very small quantities because they are toxic, persist in the environment, and build up in people's bodies.

Critics are many. They claim that these changes would slow information to communities about toxic releases and waste discharges, and prevent public health agencies and researchers from uncovering potential environmental risks.

Of all the regulated industries the chemical industry is probably the industry most burdened by TRI reporting. They claim the overall burden of the system has dramatically increased over the years. Industry wants streamlined reporting requirements to have the resources necessary to compete in the global market. All environmental reporting policies suffer this criticism in the early stages of implementation. The initial burden is always on those regulated because they are closest to the product and its environmental impact. Most U.S. environmental regulation is already based on the self-reported data of some regulated industries in some regulated activities. Environmentalists and other proponents of sustainability argue that regardless of any burden, accurate reporting of all chemicals is minimally necessary. If the industry is reluctant to accept minimal regulation, many communities do not trust them. This is often the beginning of citizen monitoring of environmental decisions. The policy crinkle in this controversy is at the beginning of implementation when the start-up costs are the greatest. The battleground for this controversy is as large as the public's desire to know about the environment.

Overall, the TRI reporting program has been a catalyst for improved accounting and transparency regarding the emissions of toxic chemicals. The benefits of the program have been a reduction in the amount of toxics released to the environment, the standardization of methods for calculating emissions, and education of the general public about the type, amount, and location of toxics released by the industry. The program has been in place since 1987, and regulated companies have adapted to the effort and burden of reporting. Moreover, the public is accustomed to receiving annual updates of toxic releases.

Other critics point to technology as part of this controversy. They point to how technologically behind the times the EPA, industry, and environmental reporting are in the United States. Companies have spent billions of dollars since the early 1990s increasing productivity through information technology. Similar tools exist for TRI reporting. These tools go beyond what the EPA discusses in its Federal Register proposal for reporting the numbers. These tools can help companies gather daily information about air, water, and waste emissions and, with the push of a button, total emissions data on a daily, weekly, monthly, quarterly, semiannual, or annual basis. These tools can extract the information from a central database to the EPA TRY-ME reporting files from where they can be electronically submitted to the EPA. If companies invested more in the proper productivity tools for environmental engineers and scientists, then reporting would not be expensive.

POTENTIAL FOR FUTURE CONTROVERSY

Environmental information is necessary for all environmental decisions. All transactions about pollution need to be transparent so that environmental information can be checked and verified. Lack of environmental information seldom stops a controversy and often inflames it. Communities do not trust industries that do not want to give information about their use of chemicals in their midst. Industries fear lawsuits and other industries getting a competitive advantage with the release of the toxics information. Environmentalists point out that the current TRI does not include all industries, that it is poorly enforced, and that much controversy exists over the listing of toxics.

See also Community Right-to-Know Laws; Permitting Industrial Emissions: Air; Permitting Industrial Emissions: Water

Web Resources

Canadian Toxic Release Inventory. Available at www.ec.gc.ca/pdb/npri/. Accessed January 22, 2008.

U.S. Environmental Protection Agency. The TRI Explorer. Available at www.epa.gov/triex plorer/. Accessed January 22, 2008.

U.S. Environmental Protection Agency. The TRI Inventory Program. Available at www.epa. gov/tri/. Accessed January 22, 2008.

Further Reading: Collin, Robert W. 2006. *The EPA: Cleaning Up America's Act*. Westport, CT: Greenwood; Hamilton, James T. 2005. *Regulation through Revelation: The Origin, Politics, and Impacts of the Toxics Release Inventory*. Cambridge: Cambridge University Press; Irwin, Frances, Tundu Lissu, Crescencia Maurer, and Elena Petkova. 1995. *A Benchmark for Reporting on Chemicals at Industrial Facilities*. Washington, DC: World Wildlife Fund.

TRANSPORTATION AND THE ENVIRONMENT

Transportation is a major contributor to air pollution, with motor vehicles accounting for a large share of nearly all the major pollutants found in the atmosphere. Trains, planes, trucks, and cars define the transportation system and have large environmental impacts. As these impacts become known and begin to accumulate in communities, many urban communities resist transportation enlargements such as roads.

The movement of people, goods, and materials requires large amounts of energy. Much of this energy is reliant on nonrenewable fuel reserves such as gas and oil. They also produce pollution that affects the land, air, and water. It is the battleground of air pollution that creates some of the most intense controversy.

Bus and railway depots in urban areas can be sinks of polluted air. These sinks will increase as these facilities expand to meet transportation demands. Increased transportation demand is reflected in longer and bigger traffic jams and gridlock. These themselves increase the amount of emissions that spew into the surrounding air. Many urban communities are already overloaded with transportation modalities that tend to benefit those outside the city. The notorious electrified third rail of the New York City subway system poses a deadly hazard wherever it is exposed. That subway tends to run underground in wealthy areas and above ground in generally lower-income and diverse communities. Many large cities east of the Mississippi have similar mass transit approaches. Mass transit has often left these communities more exposed to transportation hazards. Some maintain that rich and powerful white communities get better-designed roads with higher safety margins than poor, African American and Hispanic communities. As these emissions have accumulated, and these communities become environmentally self-empowered, the resistance to enlargements in transportation infrastructure is vigorous. Local battlegrounds may include land-use hearings, environmental impact assessments, and courts. For example, Portland, Oregon, would like to add a fifth lane to the four-lane interstate highway to accommodate commuters from the outlying, predominantly white suburbs. They want to add a lane in a Portland community that is lower income and very diverse and that already has large amounts of air pollution. These battlegrounds have been a series of community and city meetings, with the city trying to persuade the community that the land expansion is important. The neighborhoods strongly resist and do not think anything could mitigate environmental impacts enough to reduce the area's 14 percent asthma rate. Another example is in Seattle, Washington. After years of controversy and public ballots, Seattle is building a better mass transit system. Because there are significant environmental impacts the U.S. Department of Transportation had to perform an environmental impact assessment. Early plans replicated the U.S. East Coast pattern of delivering infrastructural improvement based on the wealth and race of the community. Because the environmental impact assessment did not adequately address environmental justice impacts, they had to do it all over again and make significant changes in the transit plan. The litigation and result held up about $47 million in federal aid until the environmental assessment was performed to a satisfactory level.

The environmental impacts of mass transit and private transportation are well known. Both transportation types are increasing, and communities are increasing their resistance because of the environmental impacts. There have been attempts to handle aspects of this problem with federal legislation. Although the

Intermodal Surface Transportation Efficiency Act of 1991 strongly reinforced the Clean Air Act requirements through its planning requirements and flexible funding provisions, technical uncertainties, conflicting goals, cost-effectiveness concerns, and long-established behavioral patterns make achievement of air quality standards a tremendous controversy. Techniques of estimating (and forecasting) emissions from transportation sources in specific urban areas are still controversial and generally inaccurate at the individual level. The number of monitoring stations and sites remains low, which often forces the citizens to monitor the air themselves. The lack of monitoring sites is a key issue for most U.S. environmental policy. Most industry self-reports its environmental impacts, and many industries are not even required to get any kind of permit. The more monitoring sites the more potential liability the corporation faces. Communities and environmentalists do not trust government and industry when monitoring is not allowed or is insufficient. This greatly inflames any controversy but particularly air pollution controversies around transportation's environmental impacts.

Transportation activity contributes to a range of environmental problems that affect air, land, and water—with associated impacts on human health and quality of life.

NOISE

Noise is probably the most resented form of environmental impact. Despite the money devoted to noise abatement, these measures are still limited in their effects. Numerous studies have been conducted from economic points of view, but their findings can be seen as somewhat controversial. Environmental impacts of noise can affect nesting sites for birds, migration pathways and corridors, and soil stability. Noise can also decrease property values.

SPRAWL AND CARS

U.S. cities are characterized by a separation of work and home, connected mainly by cars and some mass transit systems in denser urban and older suburban areas. The desire for a single-family detached home away from work, and increasingly away from other trips like shopping and school, requires large amounts of land and therefore has greater environmental impacts. Cars and trucks on the road today are some of the heaviest contributors to poor air quality and global warming. Illnesses such as cancer, childhood asthma, and respiratory diseases have become increasingly linked to emissions from transportation. This problem is furthered by poorly designed transportation systems that contribute to sprawl, causing freeways to become more congested and polluted. Despite improvements in technology, the average fuel economy of vehicles is less than it was in the 1980s, which also means they generate more pollution. The expansion in the production of hybrid vehicles and technological improvements in conventional vehicles could raise the fuel efficiency of new vehicles to 40 miles per gallon within a decade and 55 miles per gallon

FREEWAYS' TAINTED AIR HARMS CHILDREN'S LUNGS

Southern California contains some of the dirtiest air in the United States. There are enormous traffic problems, large polluting industries, and a rapidly increasing population. The smog can extend for hundreds of miles out in the Pacific Ocean, and hundreds of miles inland to the majestic Sierra Mountains. The public health risks extend to wherever the smog accumulates. The regional air quality control boards have been intense battlegrounds. At one time all 15 scientists in the Los Angeles air basin quit, resigned, or were terminated because of the failure to set enforceable and strong clean air standards. The issue of air pollution harm is therefore a very intense controversy.

University of Southern California (USC) researchers found in January 2007 that children living near busy highways have significant impairments in the development of their lungs. These impairments, or tears and scars in the lung tissue, can lead to respiratory problems for the rest of their lives. The 13-year study of more than 3,600 children in 12 central and southern California communities found that the damage from living within 500 yards of a freeway is about the same as that from living in communities with the highest pollution levels. For communities in high-pollution areas and living near highways there is a huge increase in risk of respiratory illness. The greatest human damage is in the airways of the lung and is normally associated with the fine particulate matter emitted by automobiles. The research is part of an ongoing study of the effects of air pollution on children's respiratory health. Previous study findings show that smog can slow lung growth, and highway proximity can increase the risk of children getting asthma.

Groups of fourth-grade students began the study, average age 10, in 1993 and 1996. The USC research team collected extensive information about each child's home, socioeconomic status, and health. Once each year, the team visited the schools and measured the children's lungs. Results from the study in 2004 indicated that children in the communities with the highest average levels of pollution suffered the greatest long-term impairment of lung function. In the new study, children who lived within 500 yards of a freeway had a 3 percent deficit in the amount of air they could exhale and a 7 percent deficit in the rate at which it could be exhaled compared with children who lived at least 1,500 yards, or nearly a mile, from a freeway. The effect was statistically independent of the overall pollution in their community. The most severe impairment was in children living near freeways in the communities with the highest average pollution. According to the USC study, those children had an average 9 percent deficit in the amount of air they could expel from the lungs. Lung impairment was smaller among those who moved farther from the freeways.

by 2020 according the Natural Resources Defense Council, an environmental organization.

ENVIRONMENTAL IMPACTS

Each major highway or other transportation project impacts the environment in a variety of ways. The most immediate negative impact on the human

HAWAII: PARADISE LOST?

Without public comment, transportation officials exempted the Hawaii Superferry from an environmental review required of projects that use federal government money and have significant environmental impacts. This has generated enormous controversy among many stakeholder groups. In September 2007 the Kauai Surfers Association protested the first day of operation and successfully blocked the Superferry's physical progress.

The first ferry is a four-story, 900-passenger, 250-car catamaran built especially for Hawaii at a shipyard in Mobile, Alabama. The second is being built. The first Superferry is to make daily trips between Honolulu and the islands of Kauai and Maui with one-way fares of $42 per person and $55 per vehicle. The second ferry would add service to the Big Island. Currently, the only regular interisland travel is by air, with one-way fares ranging from $79 to more than $100.

Environmentalists resist the Superferry because of traffic congestion, collisions with humpback whales, the spread of invasive species, and strains on limited harbor space. A recent opinion by the state Environmental Council said the Department of Transportation erred when it granted the exemption for an environmental impact review. Superferry officials argue they have exceeded environmental requirements.

Two lawsuits calling for environmental evaluations, one before the Hawaii Supreme Court and another in Maui Circuit Court, are also pending. At issue is whether the Superferry would be exempt. The law states in relevant part

> Section (b) of 11–200–8…no exemption shall apply "when the cumulative impact of planned successive actions in the same place, over time, is significant, or when an action that is normally insignificant…may be significant in a particularly sensitive environment."

The DOT previously had found that successive actions relating to a proposed intra-island ferry (on Oahu) in the early 1980s did require preparation of an environmental impact statement, and thus there was a distinct inconsistency in their application of an exemption in the case of the Superferry. Both the cumulative and secondary effects evidenced through numerous resolutions of county officials in Maui, as well as the established sensitivity of the environment in which the Superferry will operate (conservation district, shoreline, endangered species, etc.) meet the explicit terms of §11–200–8(b).

The ultimate concern here lies in the avoidance of systematic environmental review. No amount of after-the-fact study or mitigation undertaken by the Superferry reverses the failure to abide by the intent of the law that there should be public disclosure and consideration of serious environmental concerns as part of a process that is concluded prior to approval of a major project. One of the three lawsuits filed, specifically challenging the exemption, is under appeal to the Hawaii Supreme Court.

From the Hawaii Environmental Council, available at http://hawaii.gov/health/oeqc/envcouncil.html.

environment is the destruction of existing homes and businesses. Longer-term impacts include noise, air pollution, and potential loss of living quality. Wildlife and plants suffer from habitat destruction and various forms of pollution.

In addition, ecosystems suffer fragmentation; habitats and ecosystems that had worked together are divided. Migratory species may be separated into genetic islands, reducing future biodiversity and leading to local extinctions. Transportation projects may also necessitate the draining or contamination of wetlands, which are important for flood control and filtering and cleaning water. Current laws require that wetlands be reclaimed or created somewhere else. Critics claim these laws are poorly enforced and have many exemptions.

POTENTIAL FOR FUTURE CONTROVERSY

Transportation systems show little signs of abating in size and scale. Their environmental impacts have serious public health consequences and implications for sustainability. The air emissions from these systems accumulate, and more communities are now knowledgeable about some of their effects. The environmental impact statements required by many of these projects are battlegrounds of controversy. Mass transit and low-impact transportation modalities (like bicycles) are not accommodated in the United States. The current lack of integrated environmental land-use planning in the United States also prevents the development of alternative modalities on the scale necessary to reduce environmental impacts. The current healthy-community movement and policies do emphasize low-impact and healthy alternatives, and the physical design necessary for people to engage in these activities safely, but are still theoretical.

Controversies about roadway development and expansion will continue as environmental controversies.

See also Acid Rain; Automobile Energy Efficiencies; Climate Change; Environmental Impact Statements: United States; Sprawl

Web Resources

Environmental Stewardship and Transportation Infrastructure Project Reviews. Executive Order 13274. Available at www.dot.gov/execorder/13274/eo13274/index.htm. Accessed January 22, 2008.

Sightline Institute. Sprawl and Transportation: Research and Publications. Available at www.sightline.org/research/sprawl/res_pubs. Accessed January 22, 2008.

Further Reading: Chinn, L., J. Hughes, and A. Lewis. 1999. *Mitigation of the Effects of Road Construction on Sites of High Ecological Interest.* London: Thomas Telford; Forman, Richard T. T. 2002. *Road Ecology: Science and Solutions.* Washington, DC: Island Press; Frank, Lawrence D., Peter O. Engelke, and Thomas L. Schmid. 2003. *Health and Community Design: The Impact of the Built Environment on Physical Activity.* Washington, DC: Island Press; Meyer, John Robert, and Jose A. Gomez-Ibanez. 1993. *Going Private: The International Experience with Transport Privatization.* Washington, DC: Brookings Institution Press; National Research Council Transportation Research Board. 2002. *Surface Transportation Environmental Research: A Long-Term Strategy.* Washington, DC: National Academies Press; Sherwood, Bryan, David Frederick Cutler, and John Andrew Burton. 2002. *Wildlife and Roads: The Ecological Impact.* Singapore: Imperial College Press.

TRICHLOROETHYLENE (TCE) IN WATER SUPPLIES

Trichloroethylene (TCE) is regulated as a human carcinogen. Trichloroethylene is a major industrial solvent with 234,000 metric tons produced annually worldwide; it is used for degreasing and cleaning metal parts and electronic components. TCE has been used and discharged by major institutions such as prisons, hospitals, and military bases for years before and after its dangers were suspected. TCE has found its way into the environment. TCE appears to be widely distributed in the aquatic environment. Battlegrounds occur with proving TCE caused a specific cancer and with cleaning TCE out of the water supply. TCE is pervasive in many ecosystems. TCE is a probable human carcinogen but that is being reviewed and is a political battleground. Human cancers associated with TCE are liver/biliary, kidney, non-Hodgkin's lymphoma, cervical, and esophageal cancers. The cost of cleaning TCE up is very high given current technology

Fresh, clean, drinkable, and swimmable water is a presumed goal, social expectation, and policy mandate. Controversy swirls around many aspects of water.

Water quality is becoming more of a battleground as criteria are being developed for pollutants and as monitoring increases by other groups.

High-risk pollutants that pervade ecosystems are becoming more prominent. Trichloroethylene is one of those pollutants.

TCE USE

TCE is a common environmental contaminant. When a facility emits tons of TCE per year into the air of a neighborhood with a school, a public health battleground starts. In most U.S. locations no air quality or water quality data are available to determine the exposures of local residents and children to TCE. The problem and exposures could be much greater than suspected. This uncertainty inflames this controversy.

PROBLEMS OF PROOF PREVENT POLLUTERS FROM PAYING FOR DAMAGES

To scientifically demonstrate harm to children from exposure to TCE in school or the community is very difficult. There are many intervening variables that current scientific models of proof do not accommodate. Many cancers take years to develop, so children from the school would be adults before the onset of cancer. Proving harm in state and federal courts puts the burden of proof on the citizens. This burden of proof translates into very expensive litigation for an unknown period of time. Appeals, enforcement of any judgment, motions, and defendant bankruptcy can thwart even the most able-bodied cancer victim. Some have accused the industry, and sometimes the environmental regulatory agency and the insurance companies for the corporation or facility, of intentionally prolonging litigation to increase the chance of the citizen-plaintiff dying before a judgment is rendered. These types of judicial and insurance battlegrounds pervade environmental controversies around TCE.

CANCER

One of the biggest concerns from TCE is cancer. Many cancers are treatable, especially if detected early enough. Cancer itself is a dreaded disease that risks death and loss of function. An ecological study of persons exposed to TCE in drinking water supplies found incidence of leukemia was higher for females in towns with the highest exposure. Various other cancers have been reported following exposure to TCE. No studies have been reported regarding cancer in humans after skin exposure to TCE. Animal studies have linked TCE exposure to various types of cancers in rats and mice. High doses of TCE administered to rodents during long-term carcinogenicity studies resulted in liver and lung tumors in the mice, and tumors of the kidney and testes in rats. The incidence data for lung tumors in mice, together with other tumor incidences, were used by the U.S. Environmental Protection Agency (EPA) to estimate carcinogenicity potency. This cancer rating was recently withdrawn in 2006 after the chemical had been listed as carcinogenic in two prior, large, consensus-based scientific trials. Environmentalists and community activists claim the delisting occurred because TCE is so pervasive in the environment that it would be too expensive for corporations that caused it to clean it up and still make profits. Some communities with large cancer clusters, such as the farmworker community in McFarland, California, believe that the corporate shell of protection from personal and individual liability should be lifted. Corporations may cease to do business and wind down or go bankrupt. Nonetheless, the prior owners of the corporation have individual assets that could pay for the cleanup. Many citizens feel strongly that the polluter should pay for the environmental impacts of their activities, especially if it puts dangerous chemicals in the environment in large quantities. By delisting TCE as a carcinogen, many accuse, conservative politics is protecting industry through the use of a government agency. TCE still causes cancer and is still in the environment in large quantities. Citizen empowerment regulatory tools such as the Toxics Release Inventory have enabled citizens to find out about large TCE discharges. The delisting is part of the controversy with TCE.

POTENTIAL FOR FUTURE CONTROVERSY

The environmental fate of TCE has not been well documented, and considerable controversy still exists concerning its behavior in the environment. Early literature references have concluded that TCE chemicals are not metabolized by microorganisms. More recent studies, however, are split on the issue of whether TCE is biodegraded, with one research group reporting both "no appreciable anaerobic degradation" and 40 percent degradation of TCE.

Whether TCE biodegrades makes a large difference in cleaning up TCE from water supplies. The costs of cleanups are always controversial. TCE contamination is so pervasive that it may be the biggest cleanup effort yet. There is some hope that emerging technologies such as nanotechnology may be applied to the cleanup of TCE. TCE cleanups will be part of environmental and public health controversies for the near future.

See also Cumulative Emissions, Impacts, and Risks; Ecosystem Risk Assessment; Human Health Risk Assessment; Nanotechnology; Toxics Release Inventory; Water Pollution

Web Resources

Environmental Valuation and Cost-Benefit News. Empirical Cost-Benefit and Environmental Value Estimates. Available at www.envirovaluation.org/index.php?cat=89. Accessed January 22, 2008.

Rachel's Environment and Health News #370. Chemicals and Health: Part 2 and SLAPPed. December 30, 1993. Available at www.rachel.org/bulletin/pdf/Rachels_Environment_Health_News_758.pdf. Accessed January 22, 2008.

Southern Tier Superfund Sites. Available at concernedcitizens.homestead.com/Superfund_links.html. Accessed January 22, 2008.

Further Reading: Cuesta-Camacho, David E., and David E. Camacho. 1998. *Environmental Injustices, Political Struggles: Race, Class, and the Environment.* Durham, NC: Duke University Press; Goldstein, Inge F., and Martin Goldstein. 2002. *How Much Risk: A Guide to Understanding Environmental Health Hazards.* New York: Oxford University Press; McCutcheon, Steven C., Jerald L. Schnoor, and Alexander J. B. Zehnder. 2003. *Phytoremediation: Transformation and Control of Contaminants.* Hoboken, NJ: Wiley-IEEE; Spellman, Frank R., and Joanne Drinan. 2000. *The Drinking Water Handbook.* Boca Raton, FL: CRC Press.

TRUE COST PRICING IN ENVIRONMENTAL ECONOMICS

While the environmental, social, and health costs of current economic activities mount, economists do not count them. Measuring the true price, inclusive of the cost of environmental impacts, of products is more accurate because of greater understanding of science and human impacts on the environment. By charging the true price economically consumers can have a choice to have less environmental impact because they will purchase less costly products, and industry can be motivated to stop producing pollution and waste.

THE BASIC EQUATION: PROFIT PLUS COST EQUALS PRICE

In industrial capitalism the cost of a given product is roughly equal to the costs of production plus profit. The cost of producing a given product is initially borne by the producer. The manufacturer assembles the raw materials, organizes the labor, develops and finds markets and distributors, and makes the product. Costs that are outside the costs of production are called externalities. Pollution and environmental and health consequences are usually considered external to the costs of production. Yet they are costs that are delivered in expensive ways, in environmental degradation and erosion of public health.

ECONOMICS OF TRUE COST PRICING

True cost pricing (TCP) is an accounting and pricing system that includes all costs in the price of a product. This would make ecologically sound products cheaper to the consumer in terms of market price, and the demand for these products would increase. Also, various cultural and traditional industries that have been marginalized by high-impact industrial technology could compete.

Many products carry hidden environmental and social costs such as air and water pollution, deforestation, and toxic waste. These costs are created during the production, use, or disposal of the products. While the producer internalizes revenue and profits from these products, the costs are externalized to society and the natural environment. The first serious attempt at any regulation of the environment in the United States began in 1970 with the formation of the U.S. Environmental Protection Agency (EPA). Even today many industries that self-report their emissions do not reach the threshold necessary to even require any type of permit. Environmental impacts from communities and commercial activities are seldom known, reported, or regulated. Many of the laws that exist to prevent environmental and social damage are not adequately enforced. Many environmentalists contend that the environmental standards developed by the regulatory agencies like the EPA are not stringent enough to protect the environment. Weak laws that are weakly enforced allow polluters to profit at the partial expense of the health and welfare of the public and environment. In this way, externalized costs to industry equate to a subsidy paid in part by degraded environments and decreased public health. Environmentalists and impacted communities continue to seek redress and policy from the courts and legislatures. Industry argues that the need to compete with global markets requires the least expensive means of production. U.S. corporate law requires that the chief executive officer (CEO) of a company make reasonable financial decisions on behalf of shareholders, usually meaning a profit motive. This is a powerful and rigid controversy that affects all segments of society. Some are questioning the special corporate charter given by the state to an organization structured as a corporation, partnership, subsidy, or some other legally recognized form that allows the owners to escape individual liability for the debts of their business. The government represents the people and grants this immunity. They can also take it away. The Corporate Charter Revision movement adds fuel to this controversy by challenging one of the largest and most powerful stakeholders on the planet, corporate business.

True cost pricing would measure true environmental impacts in a way that is just starting to develop as public policy in the United States. Ecological risk assessment is now applied to select cleanup projects. Citizens are monitoring the environment themselves. There is the rudimentary Toxics Release Inventory. Citizen right-to-know laws have spread across the country. The level of environmental literacy and understanding has improved and expanded. This has empowered community groups and sometimes sympathetic state environmental agencies.

True cost pricing would not only account for environmental costs but also affect other controversial policy changes. Environmental taxes such as the carbon

WHAT DOES IT REALLY COST?

Cost of Aviation

Aviation gives rise to a number of adverse environmental impacts. These include aircraft noise, contributions to local air quality problems and climate change, and other negative effects on townscape, landscape, biodiversity, heritage, and water.

Air Pollution Generated by Food Transport

A 2001 Iowa State University study showed that through the conventional food delivery system, the average piece of produce in that state travels 1,494 miles to get to the consumer. Compared to Iowa-based regional and local systems, the conventional system used far more fuel and released 5 to 17 times more carbon dioxide into the atmosphere.

Cost of Industrial Farming in Great Britain

A 1999 report from Essex University calculated that British taxpayers spend up to £2.3 billion every year repairing the damage that industrial farming does to the environment and human health. The report's authors compiled figures from a variety of sources. In 1996, water companies spent £214 million removing pesticides, nitrates, and farm pathogens from drinking water. The bill for food poisoning includes an allowance for the victims' lost wages as well as the cost of their hospital treatment. The government agency English Nature calculated that restoring endangered species and wildlife habitats damaged by agriculture costs £25 million annually. The bill of £1.1 billion for air pollution and greenhouse emissions includes, for example, the cost of flood protection as a result of rising sea levels.

Cost of Obesity in the United States

According to a 2002 U.S. Surgeon General report, 61 percent of Americans are significantly overweight. This obesity generates $117 billion in annual medical bills and triggers 300,000 premature deaths each year.

Cost of Agricultural Pollution in the Water Supply

In 2000, seven people died in the town of Walkerton, Canada, when farmland runoff polluted the town's water supply with *E. coli.* The subsequent crisis cost at least $64.5 million Canadian dollars overall and individual households had to spend about $4,000 Canadian dollars on average.

Cost of Mad Cow Disease

Disruptions to the beef industry caused by mad cow scares:

Canada—$3.3 billion
United States—$6 billion
Europe—€92 billion

negative environmental impacts and increase the cost of mitigating those impacts. Some propose to integrate true cost pricing into domestic industrial policies and regulations and promote it in international trade agreements. If true cost pricing is to work consumers must be motivated to find the lowest price, it must be accessible, and consumers must have knowledge of the production process. Therefore, true cost pricing would implement product labeling to inform consumers of the total cost of the product's ingredients and manufacturing process. Another policy component is that true cost pricing may have a large short-term impact on people of lesser financial means, so that measures to mitigate these effects may need to be implemented. True cost pricing policy components in environmental impact statements, international trade agreements, government contracts, product labeling, and mitigation measures for poor people are all controversial.

DISCOUNTING THE VALUE OF THE UNKNOWN

Most economic cost analyses focus on current or annualized expenditures. Cost generally represents money or assets that must be expended in a given agency's fiscal year. The emergence of global climate change as a major environmental policy issue and the rise in social preferences for sustainability change how cost is computed. These dynamics require a consideration of costs and benefits across very long periods of time. To compare costs and benefits over time, costs in the future are discounted to the present at a particular rate. The appropriate rate of discount has been subject to much controversy. Some value must be put on future human lives. Since there cannot be an infinite number of future humans, some value must be assigned for analysis. Value is put on human life and injury every day in courtrooms. Industries consider the cost of wrongful death lawsuits part of the cost of doing business, as Lee Iacocca did in the business development of the Ford Pinto's gas tank location. While the actual value may be disputed, the value of human life is determined by insurance companies' actuaries, and regulatory determinations are made about it all the time. Nonetheless, in the environmental area this remains a developing, awkward, and quiet controversy. This is especially true in the areas of sustainability and population control.

COSTING HEALTH BENEFITS

The U.S. Environmental Protection Agency (EPA) recently promulgated regulations to reduce air pollution from heavy-duty vehicles. The estimated health benefits of reductions in ambient particulate matter (PM) concentrations associated with those regulations, based on the best available methods of benefits analysis, were estimated. The results suggest that when heavy-duty vehicle emission reductions from the regulation are fully realized in 2030, they will result in substantial, broad-scale reductions in ambient particulate matter. This will reduce the incidence of premature mortality by 8,300, chronic bronchitis by

Cost of Global Warming

Private insurers hit hard by global warming costs prepared a report in 2001 demo
that more frequent tropical cyclones, loss of land as a result of rising sea levels, and
to fishing stocks, agriculture, and water supplies amounted to a yearly bill of $304.

Cost of Air Pollution

Exposure to air pollution affects mortality rates, hospitalizations, emergency
other medical visits, problems of asthma and bronchitis, days of work and schoc
restricted-activity days, and a variety of measures of lung damage. A World Health
zation study of France, Switzerland, and Austria found that their health costs due
pollution amounted to approximately 1.7 percent of gross domestic product (GDP),
cally more than the cost of treating injuries from traffic accidents. In Canada, the pr
Ontario estimates that air pollution costs its 12 million residents at least $1 billion a
hospital admissions, emergency room visits, and worker absenteeism.

The World Bank reports that in China, home to some of the most polluted air in t
the deaths and illnesses of urban residents due to air pollution cost an estimated
of GDP.

Cost of Driving

What would it cost to drive if the price of gas and cars included air pollution; r
struction and maintenance; property taxes lost from land cleared for freeways; free
paid for by taxes; noise and vibration damage to structures; protection of petroleu
lines; sprawl and loss of transportation options; auto accidents; and congestion? A
of researchers have tried to answer this question. The Sierra Club profiled eight stuc
when averaged, estimated the true price of gas at $6.05 a gallon. As for vehicles, t
tation analysts have calculated that the external costs of driving would add $42,36
sticker price of a shiny new car, based on a 12.5 year life span.

Cost of Noise

The 1.3 million jet skis in the United States impose approximately $900 million
costs on U.S. beachgoers each year.

It is likely that many of these costs are now higher. The cost of oil and gas continu
sharply, and energy costs form a large part of these cost estimates. Even economis
that at some point the cost of a good or natural resource does not represent its va
sustainable society. When those natural resources are depleted and gone for future
tions, the amount of lost future value is impossible to econometrically forecast. Mar
cates of sustainability advocate use of the precautionary principle in these instances

tax are nascent environmental policies. True cost pricing is a basis for decisions on government projects and in environmental impact statements. If applied to environmental impact statements, true cost pricing could enlarge the

5,500, and respiratory and cardiovascular hospital admissions by 7,500. In addition, over 175,000 asthma attacks and millions of respiratory symptoms will be avoided in 2030. The economic value of these health benefits is estimated at over $65 billion. Some communities object to putting a dollar value on any life, suggesting even putting a number on it reduces its value.

True cost pricing would put a dollar value on the health costs and add it to the product. One concern is that healthy food, air, water, and land will be for the wealthy, and the poor would get the more harmful environment. So, although true cost pricing may use market-based forces to reduce environmental and public health impacts, it may also simply make human mortality and environmental degradation part of the cost of production. Environmentalists and many indigenous peoples who maintain a stewardship perspective do not embrace true cost pricing for this reason.

POTENTIAL FOR FUTURE CONTROVERSY

True cost pricing may be the free market incorporating the consumer preferences of an environmentally enlightened citizenry. Governments are often torn between two very powerful stakeholders—their citizens' health and welfare versus economic development. The stakes are high and the various interest groups are powerful and far apart in current approaches. This controversy will have many subcontroversies as new environmental policies grapple with protecting and creating a clean and safe environment under a dominant free-market regime.

As the scale and impact on the environment of human activity increase and become known, the ability to quantify the actual complete costs of many activities will increase. In economic terms, the costs of pollution will be internalized. Some argue that some impacts, like cancer in a community or a landscape tainted by radiation, are not reducible to costs. The current battleground for this controversy is in the form of new and emerging environmental legislation that incorporates the polluter-pays principle. Under this principle the facility doing the polluting pays for all costs to people and the environment. This is a popular principle but strongly resisted by industry.

See also Environmental Impact Statements: United States; Sprawl

Web Resources

U.S. Environmental Protection Agency. Water and Wastewater Pricing. Available at www. epa.gov/water/infrastructure/pricing/index.htm. Accessed January 22, 2008.

Further Reading: Collin, Robert W. 2006. *The EPA: Cleaning Up America's Act.* Westport, CT: Greenwood; Costanza, Robert, John Cumberland, Herman Daly, Robert Goodland, and Richard Norgaard. 1997. *An Introduction to Ecological Economics.* Boca Raton, FL: CRC Press; Freeman, A. Myrick III. 2003. *The Measurements of Environmental and Resource Values: Theory and Methods.* Washington, DC: Resources for the Future; Lovins, Amory, and Paul Hawkin. 1999. *Natural Capitalism: Creating the Next Industrial Revolution.* Boston: Back Bay Books.

TSUNAMI PREPARATION

Tsunamis present environmental controversies in terms of their prediction and emergency response preparation. Tourism-based coastal areas may be reluctant to warn tourists because of the potential loss of profits and revenue.

Warning communities of tsunamis is an old problem with modern implications. Warning communities can take many forms. The accuracy and quickness of the warning can affect lives. Education campaigns are apart from any warning system. Minimally, people need to know how they are going to get information in an emergency and what to do about it. Evacuation planning should also include evacuation routes. Warnings without such training are ineffective. Without evacuation planning and emergency response coordination, natural disasters can cause great harm. Ironically, without public education campaigns, untrained people may go down to the beach to see what a tsunami looks like. Public education campaigns need to target both local residents and tourists. Tourist communities may fear that such warnings could deter tourists. The incentive for local communities that rely on tourism to warn the public is small when they are in places that tsunamis would impact. Public education policies are also expensive, and many governments are reluctant to pay for them. Federal funding of such programs may start soon with new tsunami preparation programs.

Tsunamis, also known as seismic sea waves, are a series of enormous waves created by an underwater disturbance such as an earthquake, landslide, volcanic eruption, or meteorite. A tsunami can move hundreds of miles per hour in the open ocean and smash into land with waves as high as 100 feet or more.

From the area where the tsunami originates, energy waves travel outward through the water in all directions. Once the wave approaches the shore, it builds in height because the bottom pushes it up as it approaches. Often the tsunami will pull up all the water away from the coast to gain in height. The coastline and the ocean floor will shape the size and force of the wave. Tsunami waves are not predictable in sequence. There may be more than one wave, and they can change in size. A small tsunami at one coast can be a giant wave a few miles away. A tsunami can strike anywhere along the U.S. coastline. Earthquake-related movement of the ocean floor most often generates tsunamis. If a major earthquake or landslide occurs close to shore, the first wave in a series could reach the beach in a few minutes, even before a warning is issued. Most landslides that cause tsunamis are directly on the shoreline. There are several large fault lines on the West Coast of the United States that scientists are concerned will move dramatically in the next several thousand years. This is a debate, and the battleground in this context is whether warning of a tsunami would be enough because these faults are a few hundred miles offshore. Drowning is the most common cause of death associated with a tsunami. Tsunami waves and the receding water are very destructive to structures in the run-up zone. Other associated disasters are flooding, contamination of drinking water, and fires from gas lines or ruptured tanks.

TSUNAMI PREDICTION IS AN INEXACT SCIENCE

Although the geoscientists have so far failed to develop any system for predicting earthquakes, there are some methods that are applied to warn about tsunamis to protect the people of coastal areas from being washed away by the sudden surge of water. In the United States, the tsunami prediction centers still use a manual prediction system from tremor to evacuation. A correct prediction is very difficult given the current state of tsunami-monitoring systems. The cost of an incorrect prediction may be high.

In the case of Hawaiian tsunamis, three of four tsunami warnings issued since 1948 have been false alarms. It is estimated that an evacuation in Hawaii could cost as much as $68 million in lost profits and other revenue. There have been two warnings of tsunamis in Hawaii that ended in evacuations, and both were false alarms. Most times the waves do arrive. Their size, power, and frequency determine their risk to people and environment. That is very difficult to predict beacuse it depends on many factors, including the configuration of the ocean floor and the shape of the coastline. Coastline areas with substantial constriction, that is, areas that constrict water into a narrow and/or shallower space, could see much higher waves than more open coastline areas.

WHAT DOES A TSUNAMI WARNING SYSTEM COST?

United Nations organizations and concerned governments are creating a new warning system for the Indian Ocean. The lowest-cost components of a tsunami warning system are simple water-level gauges that measure immediate water movement. These cost about $5,000 to $8,000 per gauge. Siting and the level of monitoring and maintenance determine their effectiveness at giving enough warning. These gauges can cost substantially more if they have better monitoring equipment. It is possible to make monitoring stations with imaging capacities as well as the whole range of equipment to measure ocean characteristics necessary. An important instrument is the seismometer, because earthquakes and fault-line shifts can cause tsunamis. Satellite communication capacity is available. Unlike simple depth gauges, these systems can detect and communicate increasing risk of a tsunami much more quickly. They could also serve other monitoring purposes, such as measuring ocean salinity changes and deviations in water current. These systems can be very expensive, but some have argued a worldwide basic system would only cost several million U.S. dollars. On the global scale that is a small amount. There is a need for better monitoring equipment. The best standard for tsunami measurement is a new generation of deep-sea sensors. These devices sense when a tsunami passes over and transmit data to satellites, which then pass the signal along to warning centers. Only seven of these tsunameters are in use to date. They cost $250,000 apiece with annual maintenance costs of $50,000. Controversies ensue over the cost of these warning systems, as well as annual maintenance. Who pays for these systems and their maintenance? Who controls these systems? Scientists, environmentalists,

WHAT IS TSUNAMIREADY?

Through the TsunamiReady program, the National Oceanic and Atmospheric Administration's National Weather Service gives communities the skills and education needed to survive a tsunami before, during and after the event. TsunamiReady helps community leaders and emergency managers strengthen their local tsunami operations. TsunamiReady does not mean tsunami proof. TsunamiReady communities are better prepared to save lives from the onslaught of a tsunami through better planning, education and awareness. Communities have fewer fatalities and property damage if they plan before a tsunami arrives. No community is tsunami proof, but TsunamiReady can help communities save lives.

Business leaders, civic groups, political leaders, and local government officials can be instrumental in helping their community to become TsunamiReady.

To be recognized as TsunamiReady, here are some of the criteria that a community must meet.

- Establish a 24-hour warning point and emergency operations center.
- Have more than one way to receive tsunami warnings and to alert the public.
- Promote public readiness through community education and the distribution of information.
- Develop a formal tsunami plan, which includes holding emergency exercises.

and some communities advocate for them. Countries and local governments seem lukewarm to many tsunami warning systems.

Seismometers and coastal tide gauges do not provide data that allow accurate prediction of the impact of a tsunami at a particular coastal location. Monitoring earthquakes gives a good estimate of the potential for tsunami generation, based on earthquake size and location. It gives no direct information about the tsunami itself. Partly because of these data limitations, 15 of 20 tsunami warnings issued since 1946 were considered false alarms because the tsunami that arrived was too weak to cause damage.

Recently developed real-time, deep-ocean tsunami detectors provide the data necessary to make better tsunami forecasts. Amplitudes, arrival time, and periods of several first waves of the tsunami were correctly forecasted in recent scientific experiments. To many, it seems that tsunami forecasting is ready to move forward in accuracy. Initial battlegrounds of cost and control are so far preventing a unified global effort. Another reason may be that even with accurate tsunami forecasting all coastlines are unique. The shape of the land as the tsunami wave train moves in is a strong factor in determining amplitude.

POTENTIAL FOR FUTURE CONTROVERSY

The vast majority of coastal communities in the United States and around the world have inadequate tsunami warning systems. Fewer still have emergency response and evacuation plans. World and U.S. coastal populations continue to increase, rising sea levels are predicted as a result of global warming, and climate change models predict an increase in violent weather in some regions.

It is unclear who opposes early warning for tsunamis and earthquakes. Emergency preparation and evacuation planning would seem to be a high priority for fire, police, and public health workers. Without adequate warning, these on-the-scene service providers are thrown into a reactive, crisis-intervention mode of action. Without an adequate emergency response plan they do not know with whom or how to communicate what kind of information. Without public education policies that reach residents, businesses, and tourists, people do not know what to do or where to go. Without an evacuation route no one knows how to get out, especially if roads are jammed with traffic.

Tsunamis represent one of the environment's inconvenient truths. No one embraces the idea of their existence and destructiveness. No one stops tsunamis either. Scientists of all kinds from around the world clamor for better monitoring. There are recent increases in some untested tsunami warning devices. There will certainly be more tsunamis to test them, and in their path of impact will be more controversies about early warning and community preparedness.

See also Climate Change; Evacuation Planning for Natural Disasters; Floods; Global Warming

Web Resources

National Oceanic and Atmospheric Association. Preparedness and the Tsunami Resilient Community. Available at www.tsunami.noaa.gov/prepare.html. Accessed January 22, 2008.

National Tsunami Mitigation Program History. Available at http://nthmp-history.pmel.noaa.gov/. Accessed January 22, 2008.

Further Reading: Bernard, E. N. 1991. *Tsunami Hazard: A Practical Guide for Tsunami Hazard Reduction.* New York: Springer; Fredericks, Anthony D. 2002. *Tsunami Man: Learning about Killer Waves with Walter Dudley.* Honolulu: University of Hawaii Press; Hebenstreit, Gerald T. 2001. *Tsunami Research at the End of a Critical Decade.* New York: Springer; Satake, Kenji. 2005. *Tsunamis: Case Studies and Recent Developments.* New York: Springer.

W

WATER ENERGY SUPPLY

Controversial aspects of hydroelectric power involve the effects of dams on water and wildlife. Downstream water quality, stream erosion, and loss of wildlife habitat are frequent battlegrounds. Upstream issues include flooding whole communities, sedimentation, and species extinction. Dams affecting endangered fish are now being torn down in a new wave of environmental controversies associated with dams.

Running water as an energy source has been part of human civilization for thousands of years. It was primarily used to turn water wheels that turned large stones that ground grains and corn into meal. The first use of water as a source of electricity was in 1882 in the United States. Now it is one of the main sources of energy in the world, and often sought after by developing nations with the natural and financial resources. There are many hydroelectric power stations, providing around one-fifth of the world's electricity.

DAMS: THE MAIN SOURCE OF HYDROELECTRIC POWER AND CONTROVERSY

Dams are ancient ways to control and regulate water flow. With modern building technology the scale of dams has increased enormously.

Three Gorges Dam

The $24 billion Three Gorges Dam on the Yangtze River in China will be the largest hydroelectric dam in the world. It would span nearly a mile across and tower 575 feet above the world's third-longest river. Its reservoir would stretch over 350 miles upstream and force the displacement of close to 1.9 million people. (The Chinese government says 1.2 million people will be relocated, although others dispute this number being too low.) Construction began in 1994 and is scheduled for completion by 2009. Its environmental impacts are extensive and many remain unknown. China has dammed more of its rivers than most nations.

Dams have mobilized environmental protests from neighbors, environmentalists, and others for about two centuries. In the United States, property rights and water rights are in an awkward balance to this day, with differences between western and eastern U.S. water laws. Dams can have significant environmental effects. With modern construction techniques, dams can be made quite large, and the fear is that these large-scale projects can have large, global, environmental impacts.

Environmental Impacts

There are many controversies about environmental impacts. The dam builders conclude that impacts could actually enhance the environment. However, environmentalists claim that the Canadian company hired to build the dam based their assessment on a previous Chinese government environmental impact statement, which dismissed the project's environmental effects as insignificant. The Chinese impact statement failed to meet China's National Environmental Protection Agency guidelines. Environmentalists claim that the Canadian dam builders' conclusions are inadequate, misleading, and irresponsible because they neglected the social and environmental disruption that the dam would cause to the following.

- The 75 million people who live along the Yangtze River downstream of the Three Gorges whose subsistence economies are inextricably tied to the ecosystems along the Yangtze and around the downstream lakes and wetlands.
- Land-use patterns due to resettlement of environmental refugees.
- Downstream lakes and wetlands that support productive fisheries and provide critical habitat for endangered Asian waterfowl such as the Siberian crane.
- Riparian, estuarine, and marine fisheries that are already suffering a serious decline in productivity due to pollution, dammed tributaries, and overfishing.
- Coastal flooding and erosion of hundreds of square miles of China's best agricultural land.
- Increasing the salt in the city water supplies of several large cities.
- Wildlife such as the Yangtze River dolphin, the Chinese sturgeon, the finless porpoise, and the Yangtze alligator, which are already endangered and could become extinct.

The dam builders state that the most significant environmental impact would be impoundment of the river. However, others disagree with this conclusion. In addition to creating a 600-kilometer lake, impoundment of the river would flood upstream tributaries and valleys, which would effectively increase the reservoir area by 50 percent over that designated by the dam builders. The dam builders neglected to assess the social and environmental impacts of flooding upstream tributaries and their valleys. The dam builders state that flood control is the primary need for the Three Gorges project. This too is contested for the following reasons.

1. There is confusion and inconsistency in the dam, builders' study as to the area that the Three Gorges Dam would actually be able to protect from Yangtze floodwater.
2. An inaccurate method of analysis was used to determine flooding patterns downstream of the dam.
3. The dam builders failed to demonstrate that more people would be protected from flooding downstream by operation of the dam than would be flooded out as the reservoir level rises during major floods.
4. The area within the reservoir, used for flood storage in the event of a flood, would remain populated with nearly one-half million people. These people, who previously were safe from flooding, would face an increased risk of flooding.

The dam builders imply that the Three Gorges project must be built to alleviate China's energy crisis. Environmentalists again disagree with this conclusion. China has a shortage of electricity in its urban centers due to grossly inefficient energy use. The electricity needs now being met could be covered using just 60 percent of the country's existing hydroelectric capacity, leaving China with a reserve of the remaining 40 percent of hydroelectric capacity and all of its fossil fuel–burning plants. Rather than build the Three Gorges project, a more environmentally sound and cost-effective alternative would be energy-efficiency improvements and conservation measures through technological innovations and price reforms.

The Three Gorges Dam has significant potential for future controversy. In the local battleground, many early computations supporting the construction of the dam were based on incomplete data. The extensive clear-cutting of forests and global warming may affect the dam in unforeseen ways. The sedimentation caused by the dam and the effects of the dam on endangered species both promise to be a continuing controversy.

Underlying this controversy is the influence of international concern on a country's efforts to economically develop. There are many treaties on this point. China is trying to modernize its economic development and needs power to do so. Hydroelectric power is cheap and available. The scale of this modernization effort could be larger than any before it. The modern-day scale of some environmental impacts from very large projects concern international conservationists. China and other developing nations argue that the industrialized nations of the United States, Japan, and Germany had unlimited access to these power sources when they economically developed. One battleground for this controversy is the global community of environmentalists.

Another controversial area of concern represented by the Three Gorges Dam project is a fear that nonindustrialized countries will mimic the conspicuous consumption patterns of the industrialized countries. With their large and rapidly growing populations, an increase in consumption to these high levels could severely drain natural resources and promote environmental degradation.

Developing nations consider it unfair that they should be denied the same access to a higher quality of life as already enjoyed by developed nations. The battleground for this aspect of the controversy tends to be international bodies such as the United Nations, the World Health Organization, and the World Bank.

A dam is meant to hold water. All kinds of geographic, business, and political factors accompany large dam projects. In the 1960s and 1970s in the United States, environmentalists emerged as a powerful political factor. It was the environmental opposition to the Hetch Hetchy dam in southern California that ignited the late David Brower, the fiery director of several large environmental groups and projects. Canada produces the most hydropower in the world. The United States produces the second most hydropower. It is considered very clean because it does not produce any greenhouse gases or other air pollution and also no hazardous waste products.

The best site for a dam is usually in a valley where there is an existing lake or natural reservoir. There is more potent ional gravitational energy to drive turbines. At the bottom of the dam there are tunnels that can be controlled. At the bottom of the tunnels are the massive turbines. Water flows through tunnels in the dam, turns the turbines, and drives generators that create electricity that is sent to power stations for distribution to the grid. The water then reenters the river and continues downstream. Since the power station can control the tunnel gates, they can control the flow of the water downstream. Many factors affect the decisions about the tunnel gates, some environmental, some market, and some regulatory. The rate of the flow is a very important environmental issue for everyone downstream. Kayakers like to ride the release, the first wave of water

---

OCEAN POWER?

Researchers around the world have experimented with the use of tides to produce energy. By putting buoys that float on the tide as it moves in and out, different experimental technologies are seeking to obtain electrical power from the oceans.

Reliability of energy source is an important part of current research into alternative energy sources. Tidal power was thought to be unreliable because tides can differ by season, day, and locale. Tides are generally predictable, however. Even when tides are not predictable, better monitoring systems and satellite imagery of the ocean now allow for some control over the unreliability of tides. There may be other technological challenges current research is now exploring.

There will be controversy. Commercial and sport fisherman, surfers and recreational boaters, ocean shipping lanes, and others all claim some use of the coastline. Depending on where the buoys are located, and the technology of power transmission, there could be conflict over the use of that space with these groups. There could also be controversy over the direct and indirect environmental impacts. That would depend on many speculative and site-specific factors now, such as location, presence of sensitive species or marine ecosystems, or pollution.

---

after the tunnels are opened. Other boaters, people who fish, indigenous people, and industrial, agricultural, and municipal users of water all have a claim on part of that water flow, some of which are legal claims. The environmental impacts are huge with water flow. Flow determines many other factors important for aquatic life, such as salinity, turbidity, temperature, and water depth.

Hydroelectric dams are very expensive to build. Water exerts a powerful and persistent force, requiring a large amount of rebar and concrete to hold it back. It requires good roads to the site to build a dam. Once built, a power station and grid are necessary, and they also require built infrastructure. In mountainous countries such as Switzerland, hydroelectric power provides more than half of the country's energy needs. Parts of Africa and South America have vast untapped resources for hydroelectric capacity. A less expensive, less powerful alternative to a dam is to build the power station next to a fast-flowing river with enough reliable flow of water. The flow of the water cannot be controlled, which limits the control of power output. This decreases reliability of the electrical grid.

BATTLEGROUNDS

Hydroelectric power from dams is littered with battlegrounds. Some dams are being torn down, some voluntarily, some pursuant to court order, and some pursuant to environmental impact statements. Dams are very expensive to build, which translates into higher electricity costs for ratepayers later. Dams have other uses than power generation. They are used for flood control, irrigation, recreation, and fire safety. Building a large dam almost always floods a very large land area upstream. It completely destroys most life that cannot move from that area. Another basic battleground is finding a suitable site. The environmental impact of the dam on residents and the environment may be unacceptable. Water flow, quality, and quantity downstream can be affected, which can have an impact on fish and plant life. Endangered species, tribal rights, and interstate compacts may also be affected. Municipal and agricultural users of water also have concerns.

FISH LADDERS: ENOUGH TO PREVENT EXTINCTION?

Many fish are driven to return to their place of birth to spawn, after which they often die. The next generation of fish repeats this cycle as they have for many years. If their access to their birthplace is blocked, spawning does not occur and the species goes extinct. This biological fact supports the governmental provision of fish, through fish hatcheries. Wild salmon are often protected as an endangered species, but hatchery salmon are not. An intense battleground concerning wild versus nonwild salmon is opening up in supermarkets, fishing camps, and the federal courts handling endangered species cases. The fish ladder allows wild salmon to migrate upstream to mitigate the biological impact of the dam. These are very controversial in terms of their effectiveness. They do help mitigate extensive damage to fish populations, especially fish that migrate upriver. Some dams now videotape every single fish to ensure accurate fish counts.

ROOTS OF DAM CONTROVERSIES: PRIVATE PROPERTY AND CONTROL OF NATURE

Dams have an enormous impact on the environment because they fundamentally alter the watershed in a given ecosystem. Many watersheds are not necessarily stable, with periods of dry and wet, changing river courses, and different rates of eutrophication in its lakes and ponds. Damming the water affects those who live downstream and modifies the habitats of plant and animal life in the area. Private property law has recognized rights to the use of water bordering one's land (called *riparian rights*). This includes not having it diverted away from one's land by upstream dams, as a general principle of common law. However, there are many formal and informal arrangements about water use in the West that differ. Building a dam is a huge imposition on private property rights. The government can take the land via eminent domain. This will increase the intensity of the battleground.

Dams have many purposes. They are built for controlling floods, generating power, providing drinking water, improving navigation, and recreational, industrial, and agricultural uses. At the beginning of the nineteenth century, hydro-

---

DAVID R. BROWER (1912–2000)

David Brower was a leader of the U.S. environmental movement. He was a serious mountaineer who advocated for the protection of the environment and the first executive director of the influential and powerful Sierra Club, serving from 1952 until 1969 and then on their board of directors periodically until the late 1990s. He also founded Friends of the Earth, a 68-country global environmental group, and cofounded the League of Conservation Voters. This national organization closely tracks how elected officials vote when environmental issues are at stake. Sometimes they issue "report cards" or "scorecards" on how various elected officials are doing with regard to environmental voting. These organizations came at a crucial time in U.S. environmental history. The U.S. Environmental Protection Agency was freshly formed in 1970, and about 18 laws went into effect over the next ten years. Part of the success of establishing a national system of environmental protection, when there was none, is due to David Brower and his unflagging sense of purpose. Later, he founded the Earth Island Institute to empower environmental activists. In 1999 he cofounded the Alliance for Sustainable Jobs and the Environment.

His environmental activism is legendary. He led thousands of people in the Sierra Club Outings program into wilderness areas from 1939 to 1956. He stopped major dam projects in the West, he advocated for and achieved the creation of new national parks, and he was crucial to the passage of the Wilderness Act of 1964. He was not afraid of controversy. His vision protected untold acres of pristine wilderness for future generations. David R. Brower died in 2000 in Berkeley, California.

electric power from dams supplied about a third of the generating capacity in the United States. It is currently used to supply only about 10 percent of the U.S. energy supply. According to the Army Corps of Engineers the most common primary benefit for which dams are built is recreation (35%). Only 11 percent of dams were constructed for the primary purpose of irrigation, and 2 percent primarily for hydroelectric power.

POTENTIAL FOR FUTURE CONTROVERSY

Dams are part of the all-encompassing controversies around freshwater. Tapping the energy of falling water will become more attractive to users of energy as nonrenewable sources become depleted. However there are many other users of the water, a decreasing resource and commodity. Advances in technology and ecological understanding of marine systems may help better mitigate impacts. Currently, the first dams are being taken down. Many are considered to have serious environmental impacts to endangered species. The utility companies offer staunch resistance to dam decommissioning, threatening that the ratepayers will have to pay more for their power. Many proponents of self-sufficiency and sustainability advocate for these small hydrosystems if the terrain is appropriate. The controversy over the use of water for power via dams shows no signs of abating. The battleground for these controversies will be the courts.

See also Ecosystem Risk Assessment; Endangered Species; Floods; Sacred Sites; Sustainability

Web Resources

Energy Kids Page. Available at www.eia.doe.gov/kids. Accessed January 22, 2008.

Environmental Defense Fund—70 Related Documents. Available at http://www.environ mentaldefense.org/home.cfm. Accessed January 22, 2008.

Three Gorges Probe. Available at http://www.threegorgesprobe.org/tgp/index.cfm. Accessed March 2, 2008.

Turning to Hydropower. Available at www.fwee.org. Accessed March 2, 2008.

U.S. Geological Service. Water Science for Schools. Available at ga.water.usgs.gov/edu/wuhy.html. Accessed March 2, 2008.

Further Reading: Chasek, Pamela S. 2000. *The Global Environment in the Twenty-First Century: Prospects for International Cooperation.* New York: United Nations University Press; Economy, Elizabeth C. 2004. *The River Runs Black: The Environmental Challenge to China's Future.* Ithaca, NY: Cornell University Press; El-Ashry, Mohamed T., and Diana C. Gibbons. 1988. *Water and Arid Lands of the Western United States: A World Resources Institute Book.* Cambridge: Cambridge University Press; Espeland, Wendy Nelson. 1998. *The Struggle for Water: Politics, Rationality, and Identity in the American Southwest.* Chicago: University of Chicago Press; Heggelund, Gjørild. 2004. *Environment and Resettlement Politics in China: The Three Gorges Project.* Aldershot, UK: Ashgate Publishing; Khagram, Sanjeev. 2004. *Dams and Development: Transnational Struggles for Water and Power.* Ithaca, NY: Cornell University Press; Murray, Geoffrey, and Ian G. Cook. 2002. *Green China: Seeking Ecological Alternatives.* London: Routledge; Newson,

Malcolm David. 1997. *Land, Water, and Development: Sustainable Management of River Basin Systems.* London: Routledge; Postel, Sandra. 1997. *Last Oasis: Facing Water Scarcity.* New York: W. W. Norton and Company; Shiva, Vandana. 2002. *Water Wars: Privatization, Pollution and Profit.* London: Pluto Press.

WATER POLLUTION

Water is essential for life. Water quality is often at odds with the demands of increased development. Conflicting laws, poorly enforced environmental regulations, and increased citizen monitoring are the ingredients for powerful and long-lasting controversy.

Water pollution is a term that describes any adverse environmental effect on water bodies (lakes, rivers, the sea, groundwater) caused by the actions of humankind. Although natural phenomena such as volcanoes, storms, and earthquakes also cause major changes in water chemistry and the ecological status of water, these are not pollution. Water pollution has many causes and characteristics. Humans and livestock produce bodily wastes that enter rivers, lakes, oceans, and other surface waters. These wastes increase the solids suspended in the water and the concentration of bacteria and viruses, leading to potential health impacts. Increases in nutrient loading may lead to eutrophication, or dead zones, in lakes and coastal water. Organic wastes deplete the water of oxygen, which potentially has severe impacts on the whole ecosystem. Industries and municipalities discharge pollutants, permitted and sometimes unpermitted, into their wastewater, including heavy metals, organic toxins, oils, pesticides, fertilizers, and solids. Discharges can also have direct and indirect thermal effects, especially those from nuclear power stations, and also reduce the available oxygen. Human activities that disturb the land can lead to silt running off the land into the waterways. This silt can have environmentally detrimental effects even if it does not contain pollution. Silt-bearing runoff comes from many activities including construction, logging, mining, and farming. It can kill aquatic and other types of life. Salmon, for example, do not spawn if the temperature of the water is too high.

Another environmental controversy around water quality is that when water becomes polluted, native species of plants and animals fail to flourish in rivers, lakes, and coastal waters. Depending on how exactly the water quality is impaired, some of these species may be threatened with extinction. If, for example, the water quality is impaired through agricultural runoff containing nitrogen and other chemical fertilizers, this may precipitate algae blooms. These blooms can warm up the water as well as rapidly deplete oxygen in the water.

Pollutants in water include chemicals, pathogens, and hazardous wastes. Many of the chemical substances are toxic. Many of the municipal water supplies in developed and undeveloped countries can present health risks. Water quality standards consist of three elements: the designated uses assigned to those waters (such as public water supply, recreation, or shellfish harvesting), criteria to protect those uses (such as chemical-specific thresholds that should not be exceeded), and an antidegradation policy intended to keep waters that do meet standards from deteriorating from their current condition.

Water regulations control point sources of pollution. Some environmentalists consider the definition of point source too narrow because it allows smaller discharges into the water. It has been estimated that between 50 and 80 percent of water pollution comes from nonpoint sources. Nonpoint source (NPS) pollution comes from many sources, including human habitation and industrial emissions currently unaccounted for. NPS pollution begins with precipitation moving on and through the ground. As the force of gravity pulls the water down, it carries with it natural and human-made pollutants. Many of these pollutants end up in lakes, rivers, wetlands, coastal waters, and underground sources of drinking water.

These pollutants include:

- Excess fertilizers, herbicides, and insecticides from agricultural lands and residential areas;
- Oil, grease, and toxic chemicals from urban runoff and energy production;
- Sediment from improperly managed construction sites, crop and forest lands, and eroding stream banks;
- Salt from irrigation practices and acid drainage from abandoned mines; and
- Bacteria and nutrients from livestock, pet wastes, and faulty septic systems.

Atmospheric deposition is also a source of nonpoint source pollution. An incinerator next to a lake could be a source of water pollution.

NONPOINT SOURCES

States report that nonpoint source pollution is the leading remaining cause of water quality problems. The effects of nonpoint source pollutants on specific waters vary and may not always be fully assessed. These pollutants have harmful effects on drinking water supplies, recreation, fisheries, and wildlife. With only about 20 percent of lakes and rivers being monitored in any way, and much to learn about the movement of underground water and aquifers, the degree of uncertainty as to nonpoint sources is currently very large. Even water areas that are monitored still allow permits to industries and cities to discharge treated and untreated waste and chemicals.

Nonpoint source pollution results from a wide variety of human activities on the land. These activities touch upon battlegrounds in areas from private property to corporate environmental responsibility. Governmental responses to water pollution from nonpoint sources are spread across the spectrum. Some activities are federal responsibilities, such as ensuring that federal lands are properly managed to reduce soil erosion. Some are state responsibilities, for example, developing legislation to govern mining and logging and to protect groundwater. Others are local, such as land-use controls like erosion control ordinances. The coordination of intergovernmental relations and communication between these levels of government about water pollution approaches are poor, contributing to the controversy.

The United States developed new environmental policies in the past 35 years to clean up water pollution by controlling emissions from industries and municipal sewage treatment plants. This last 35-year period was preceded by 500 years of urbanization and then industrialization and waves of immigration from every coast. There was little in the way of enforceable environmental legislation in the United States until 1970. Navigable waterways have been intentionally and unintentionally altered in drastic ways, such as channelizing the Mississippi River by the U.S. Army Corp of Engineers. Modern plumbing devices, such as backflow regulators, help keep wastewater separate from drinking water. Urbanized areas without backflow regulators on industry eventually taint the entire watershed. In many areas, it is often the case that as water quantity goes down so does water quality. In places such as Texas that practiced a form of waste discharge called deep well injection, some of the water sources may be contaminated. The accumulated wastes from the water pollution both before and after the formation of the EPA will themselves foster cleanup controversies. Fear of liability for past acts of environmental contamination is a powerful contour in the battleground of water pollutant cleanup.

Nonpoint source pollution is the largest source of water quality problems in the United States. It remains the catchall term for all other than point sources of water pollution. Point sources are regulated by the EPA. Each watershed is allowed to have a limited overall amount of water pollution permits. If all sources were counted, including nonpoint sources, the overall amount of permissible chemical discharges into the watershed would decrease. This could result in fewer permits being issued. The fewer permits issued generally means less industrial economic development. Industries and governments prefer more industrial and manufacturing economic development. Some industries prefer not to compete with other industries in the same watershed and may not want to share a water permit. Uncertainty of the water permit can deter financial investors from long-term investments in a plant or real property. Other stakeholders like farmers, agribusiness, and Native Americans all hold various rights and expectations for the same water. Nonpoint sources of water pollution have serious unresolved environmental issues that involve many stakeholders. Accurate environmental monitoring is a necessity as the foundation of sound environmental policy, especially if sustainability is the goal. The range of disrespect for the environment from some stakeholders shocks other stakeholders, who feel reverence for the environment when it comes to water pollution. The wide range of environmental expectations becomes controversial when accurate environmental monitoring and research reveal the true extent of the environmental impacts of water pollution.

KNOWN SOURCES OF NONPOINT WATER POLLUTION

Agribusiness is the leading source of water quality impairments, degrading 60 percent of the impaired river miles and half of the impaired lake acreage surveyed by states, territories, and tribes. Runoff from urban areas is also a very large source of water quality impairments. Roads, parking lots, airports, and other impervious paved surfaces that occur with U.S. land development increase the runoff of precipitation into other parts of the watershed. The most common NPS pollutants are soil sediment and chemical nutrients. Other NPS pollutants include pesticides, pathogens (bacteria and viruses), salts, oil, grease, toxic chemicals, and heavy metals.

ROLE OF COMMUNITIES

Communities play an important role in addressing NPS pollution. When coordinated with federal, state, and local environmental programs and initiatives, community-based NPS control efforts can be highly successful. More than 500 active volunteer monitoring groups currently operate throughout the United States. Monitoring groups may also have information about other NPS pollution projects, such as beach cleanups, stream walks, and restoration activities. More than 40 states now have some type of program to help communities conserve water. NPS pollution starts at the household level. Households, for example, can water lawns during cooler hours of the day, limit fertilizer and pesticide applications, and properly store chemicals to reduce runoff and keep runoff clean. Pet wastes, a significant source of nutrient contamination, should be disposed of properly. Communities can also replace impervious surfaces with more porous surfaces.

RUNOFF FROM URBAN AREAS

The nonpoint sources of pollution often come from paved, impermeable road surfaces. These can be in urban, suburban, or rural areas. Many vehicle emissions run off from the pavement with water when it rains. Effective drainage systems can remove this water to city water systems, but these do not necessarily treat the runoff for its load of pollutants. In many cities, the consolidated sewer overflow system, means that when it rains heavily the sewers simply overflow into the nearest river or lake. Many urban sewer and water systems are old and need repair, especially those made with lead pipes.

Cities with storm sewer systems that quickly channel runoff from roads and other impervious surfaces increase their environmental impacts with large flow variations. Runoff gathers speed in the storm sewer system. When it leaves the system, large volumes of quickly flowing runoff erode riparian areas and alter stream channels. Native fish, amphibians, and plants cannot live in urban streams impacted by urban runoff.

POTENTIAL FOR FUTURE CONTROVERSY

Water pollution will become more controversial. As water pollution standards mature, environmental impact assessment and pollution accountability will increase. Many stakeholders now assume they have the right to fresh, clean water, and as much of it as they want. Where the water begins to run out, violent confrontations can occur. In Klamath, Oregon, the site of a furious water controversy between farmers, various agencies of the federal and state government, and environmentalists, violence erupted in 2006 as Native American children were assaulted in their school bus by farmers angry at their loss of water. Although the Klamath tribe tried to avoid the controversy, they do have water rights by treaty and law. The farmers' property rights lawsuit, claiming they owned the water as a property right, was dismissed in a 57-page opinion in federal court. Vice President Cheney is currently being investigated as illegally intervening in this dispute and commanding the federal agencies to let agribusiness get the water. This controversy only gets larger, and so far is only resolved by more water from nature.

As more and more of the aquatic environment becomes known, battlegrounds of who pays for cleanup and for dredging occur. There are battlegrounds with the environmental impacts of these activities alone. In arid developing areas where water can become scarce, those who use it and pollute it affect many other groups. There will be an increase in stakeholder accountability for pollution sources as environmental law enforcement works its way upstream. Litigation and community engagement will increase in the controversy over water pollution.

See also Citizen Monitoring of Environmental Decisions; Permitting Industrial Emissions: Water; Total Maximum Daily Loads (TMDL) of Chemicals in Water; Trichloroethylene (TCE) in Water Supplies.

Web Resources

Natural Resources Defense Council. Issues: Water. Available at www.nrdc.org/water/pollution/default.asp. Accessed January 22, 2008.

Sources of Water Pollution. Available at www.soest.hawaii.edu/GG/ASK/waterpol3.html. Accessed January 22, 2008.

U.S. Environmental Protection Agency. Polluted Runoff (Nonpoint Source Pollution). Available at www.epa.gov/nps/. Accessed January 22, 2008.

Further Reading: Best, Gerry. 2000. *Environmental Pollution Studies.* Liverpool, UK: Liverpool University Press; Helmer, Richard. 1998. *Water Pollution Control.* London: Spon Press; Houck, Oliver A. 1999. *The Clean Water Act TMDL Program: Law, Policy, and Implementation.* Washington, DC: Environmental Law Institute; Pagenkopf, James R., Andrew Stoddard, James R. Pagenkopf, and Jon, B. Harcum. 2002. *Municipal Wastewater Treatment.* New York: John Wiley and Sons.

WATERSHED PROTECTION AND SOIL CONSERVATION

Watersheds left unprotected from development may experience environmental impacts that deplete soil resources. Logging, grazing, some types of mining, paving over land with impervious surfaces, and overchannelization of major water courses have affected watersheds in controversial ways. Agriculture and environmentalists want to prevent soil depletion by protecting watersheds. Communities want to protect watersheds for water quality. For watershed protection to work as a policy it may require the taking of private property or the terminations of long-term leases given to loggers, ranchers, and mining corporations. Water use and quality are generally becoming controversial, and watershed protection is increasingly seen by some as excessive government intervention. Others see it as a necessary component of any successful sustainability program or policy.

Visions of hurricanes and floods tearing the hard earned top soil from the Midwest prompted many to ask the federal government to intervene in the 1930s. The early legislation set the tone for today's policy. Most of the federal legislation for watershed protection emerged in this time to protect rural and agricultural interests. The Watershed Protection and Flood Prevention Act of 1954, as amended, authorized Natural Resources Conservation Service (NRCS) to cooperate with states and local agencies to carry out works of improvement for soil conservation and for other purposes including flood prevention; conservation, development, utilization, and disposal of water; and conservation and proper utilization of land. NRCS implements the Watershed Protection and Flood Prevention Act through the following programs:

Watershed Surveys and Planning
Watershed Protection and Flood Prevention Operations
Watershed Rehabilitation
Watershed Surveys and Planning

The NRCS cooperates with other federal, state, and local agencies in making investigations and surveys of river basins as a basis for the development of coordinated water resource programs, floodplain management studies, and flood insurance studies. NRCS also assists public sponsors to develop watershed plans. The focus of these plans is to identify solutions that use conservation practices, including nonstructural measures, to solve problems. Each project must contain benefits directly related to agriculture, including rural communities that account for at least 20 percent of the total benefits of the project.

WATERSHED OPERATIONS

Watershed Operations is a voluntary program that provides assistance to local organizations sponsoring authorized watershed projects, planned and approved under the authority of the Watershed Protection and Flood Prevention Act of 1954, and 11 designated watersheds authorized by the Flood Control Act of 1944. NRCS provides technical and financial assistance to states, local governments, and tribes (project sponsors) to implement authorized watershed project plans for the purpose of watershed protection; flood mitigation; water quality improvements; soil erosion reduction; rural, municipal, and industrial water supply; irrigation water management; sediment control; fish and wildlife enhancement; and wetlands and wetland-function creation and restoration. There are over 1,500 active or completed watershed projects. As communities become more involved in environmental issues they quickly learn about their particular watershed.

FLOOD PREVENTION PROGRAM

The Flood Control Act of December 22, 1944, authorized the Secretary of Agriculture to install watershed improvement measures (http://www.nrcs.usda.

gov/programs/watershed/pl534.html). This act authorized 11 flood prevention watersheds. The NRCS and the Forest Service (FS) carry out this responsibility with assistance from other bureaus and agencies within and outside USDA. Watershed protection and flood prevention work currently under way in small upstream watersheds all over the United States sprang from the exploratory flood prevention work authorized by the Flood Control Act of 1944, and from the intervening 54 pilot watershed projects authorized by the Agriculture Appropriation Act of 1953. These projects are the focus of much study as watershed protection and soil conservation have become battlegrounds after the impact of Hurricane Katrina on New Orleans in 2006. Many accuse these types of projects as too little too late for prevention of risk to urban areas from natural disasters.

The 11 watershed areas are:

WATERSHED NAME	STATE	WATERSHED SIZE
Buffalo Creek	New York	279,680 acres
Middle Colorado River	Texas	4,613,120 acres
Coosa River	Georgia, Tennessee	1,339,400 acres
Little Sioux River	Iowa	1,740,800 acres
Little Tallahatchie River	Mississippi	963,977 acres
Los Angeles River	California	563,977 acres
Potomac River	Virginia, W. Virginia, Maryland, Pennsylvania	4,205,400 acres
Santa Ynez River	California	576,000 acres
Trinity River	Texas	8,424,260 acres
Washita River	Oklahoma, Texas	5,095,040 acres
Yazoo River	Mississippi	3,942,197 acres

Because the authorized flood prevention projects include relatively large areas, work plans are developed on a subwatershed basis. Surveys and investigations are made and detailed designs, specifications, and engineering cost estimates are prepared for construction of structural measures. Areas where sponsors need to obtain land rights, easements, and rights-of-way are delineated. This can present a battleground when private property owners do not want to cooperate with flood prevention and soil conservation. There are presently over 1,600 projects in operation.

WATERSHED PROJECTS PROVIDE THOUSANDS OF ACRES OF FISH AND WILDLIFE HABITAT

There are 2,000 NRCS-assisted watershed projects in the United States, with at least one project in every state. Some projects provide flood control, while others include conservation practices that address a myriad of natural resource issues such as water quality, soil erosion, animal waste management, irrigation, water management, water supplies, and recreation. Whatever the primary purpose, watershed projects have many community benefits such as fish and wildlife habitat enhancement.

Over 300,000 acres of surface water have been created by the construction of 11,000 watershed dams.

Lakes generally range in size from 20 to 40 surface acres and provide a good mix of deep water and shoreline riparian areas. Some lakes have up to several hundred acres of surface water, and many had recreational areas developed around them. Lakes formed by the watershed dams have created thousands of acres of open water providing excellent fish and wildlife habitat and areas for migrating waterfowl to rest and feed. Conservation practices in watershed projects such as buffers, pasture and rangeland management, tree plantings, ponds, conservation cropping systems, and conservation tillage provide cover, water, and food for a variety of birds and animals.

Thousands of people enjoy fishing, hiking, boating, and viewing wildlife in these very scenic settings each year. NRCS-assisted watershed projects provide a wide diversity of upland habitat landowners in watershed projects with technical and sometimes financial assistance in applying conservation practices. Many of these practices create or improve wildlife habitat and protect water quality in streams and lakes.

While watershed projects may offer benefits to recreational users, others wonder about their environmental impacts. To what end is the soil being conserved? What if the area is a natural floodplain? What are the impacts of recreational users on endangered or threatened species? These questions and others abound in the traditional soil conservation and floodplain protection policies.

CREATING AND PROTECTING WETLANDS: WATERSHED PROGRAM RESULTS

According to the Natural Resources Conservation Service, they have assisted in creating the following.

- Upland wildlife habitat created or enhanced: 9,140,741 acres
- Wetlands created or enhanced: 210,865 acres
- Stream corridors enhanced: 25,093 miles
- Reduced sedimentation: 49,983,696 tons/year

The 2,000 watershed projects have established a $15 billion national infrastructure, by their own estimate that is providing multiple benefits to over 48 million people.

- Agricultural flood damage reduction: $266 million
- Nonagricultural flood damage reduction: $381 million
- Agricultural benefits (nonflood): $303 million
- Nonagricultural benefits (nonflood): $572 million
- Total monetary benefits: $1.522 billion

- Number of bridges benefited: 56,787
- Number of farms and ranches benefited: 154,304
- Number of businesses benefited: 46,464
- Number of public facilities benefited: 3,588
- Acres of wetlands created or enhanced: 210,865
- Acres of upland wildlife habitat created or enhanced: 9,140,741
- Miles of streams with improved water quality: 25,093
- Number of domestic water supplies benefited: 27,685
- Reduced soil erosion (tons/year): 89,343,55
- Tons of animal waste properly managed: 3,910,10
- Reduced sedimentation (tons/year): 49,983,696
- Water conserved (acre feet/year): 1,763,472

The Watershed Program has been used by communities for over 50 years. The authorizing legislation has been amended several times to address a broader

WATERSHED PROTECTION AND SOIL CONSERVATION

States are important stakeholders in implementing watershed protection programs. Part of implementation is making information about the program accessible to the public.

State Watershed Web Pages

The following states have information about their watersheds available online:

Arizona	State Proclamations
California	
Colorado	Arizona
Connecticut	Hawaii
Florida	Kansas
Hawaii	Oklahoma
Iowa	
Kansas	
Louisiana	
Maine	
Minnesota	
Missouri	
Nebraska	
New York	
North Dakota	
Pennsylvania	
Utah	
Virginia	
Wisconsin	
Wyoming	

range of natural resource and environmental issues, and today the program offers communities more assistance to address some environmental issues. There are watershed projects in every state. Over 2,000 projects have been implemented since 1948. New projects are being developed each year by local people.

POTENTIAL FOR FUTURE CONTROVERSY

As water resources become scarce, the competition for water will force water users to exert all rights in water and the land. In places where floods occur, property owners will want flood control as watershed protection. Communities want clean and safe drinking water, and this is becoming a scarce resource even in communities with water. Watersheds absorb all past and present wastes, by-products, emissions, discharges, runoff, and other environmental impacts. Their protection will require increasingly stringent controls on all water users and residents of the watershed. Private property owners and land developers object to this because they may have planned for a more profitable use. Without water it is difficult to make a profit in developing land. Sustainable sources of safe water are essential to many stakeholders. Watershed protection will become increasingly controversial as it moves from 1940s and 1950s soil erosion prevention policy to one that incorporates accurate monitoring of water and precepts of sustainability in urban and suburban settlements as well as rural ones.

See also Logging; Mining of Natural Resources; Permitting Industrial Emissions: Air; Permitting Industrial Emissions: Water; Pesticides; Sprawl; Sustainability; Water Pollution

Web Resources

U.S. Environmental Protection Agency. Monitoring and Assessing Water Quality: Volunteer Monitoring. Available at www.epa.gov/owow/monitoring/vol.html. Accessed January 22, 2008.

U.S. Environmental Protection Agency. Watersheds. Available at www.epa.gov/owow/water shed/. Accessed January 22, 2008.

Further Reading: Hirt, Paul W. 1994. *A Conspiracy of Optimism: Management of the National Forests since World War Two.* Lincoln: University of Nebraska Press; Mastney, Lisa, and Sandra Postel. 2005. *Liquid Assets: The Critical Need to Safeguard Freshwater Ecosystems.* Washington, DC: Worldwatch Institute; Morandi, Larry, Barbara Foster, and Jeff Dale. 1998. *Watershed Protection: The Legislative Role.* Washington, DC: National Conference of State Governors; Riley, Ann Lawrence. 1998. *Restoring Streams in Cities: A Guide for Planners, Policymakers, and Citizens.* Washington, DC: Island Press; Sabatier, Paul A. 2005. *Swimming Upstream: Collaborative Approaches to Watershed Management.* Cambridge, MA: MIT Press.

WILD ANIMAL REINTRODUCTION

To prevent wolves from extinction, the National Park Service reintroduced them on park lands. Nearby ranchers protested this, claiming the wolves prey on their herds. Environmentalists claim ranchers killed some of the wolves. Grizzly bears may be reintroduced in national parks next.

One of the biggest reasons for the reintroduction of wolves back into Yellowstone was that this was part of their original habitat. Wolves had originally roamed from Yellowstone all the way down to Mexico. While many environmentalists and wildlife agencies were in favor of the reintroduction of the wolves, many other groups were against it. The main people who were against the reintroduction of the wolves were the ranchers who made a living in the areas surrounding the park. During its 70 years of absence from the Rockies, the grey wolf had been protected under the Endangered Species Act, which was passed in 1973. Therefore, a person could be punished with up to a $100,000 fine and up to one year in jail for killing a wolf. Back in the 1850s there was a major population increase of the wolves in the United States; this was due to settlers moving west. These settlers killed more than 80 million bison, and the wolves started to scavenge on the carcasses left behind. By the 1880s the majority of the bison were gone, so the wolves had to change food sources. This meant that they turned their attention to domestic livestock, causing farmers and ranchers to develop bounties and other vermin-eradication efforts. Due to the lack of a food source, as well as the bounties being offered, the wolf population plummeted in the lower 48 states.

When the numbers in animal population become low, the genetic diversity of that species decreases dramatically. This could hasten the extinction of a species. One of the premises of governmental intervention in these environmental controversies around species reintroduction is to prevent extinction. An important aspect of this is gene pool diversity. This requires active animal management by humans, including an in-depth knowledge of the genes of specific animals and packs. One aspect of the battleground of this controversy is the need for more scientific monitoring of animals facing extinction.

When the wolf population dropped, there was a safe place then. That was Yellowstone National Park, established in 1872. In 1916 the National Park Service started to eliminate all predators in Yellowstone National Park, which meant killing 136 wolves, 13,000 coyotes, and every single mountain lion. By 1939 this program was shut down, but all the wolves were dead.

WHAT IS A WOLF PACK?

A wolf pack is very hierarchical and organized. Dominance and submission establish the order of power in the pack. A pack consists of an alpha male, an alpha female, and their descendants. The alpha pair are the only two that breed. The natural pattern of breeding within a wolf pack works to protect their genetic diversity if populations are healthy.

THE REINTRODUCTION PROGRAM

In January 1995, 14 wolves from many separate packs were captured in Canada. They were brought into Yellowstone Park. The next step in their reintroduction was to place them in one-acre acclimation pens. Capturing wolves from different packs helped protect their genetic diversity. Biologists then created packs from these captured wolves.

While in captivity the wolves were fed large amounts of meat in the form of roadkill or winter carrion from the area. This was often deer, elk, and smaller animals. Each pack was fed once every 7 to 10 days, which is how frequently they eat in the wild. A small wolf pack of six in the wild will consume on average 800 pounds of meat per month. In Yellowstone National Park that would average out to two adult elk and maybe a small deer per small pack per month. Today, 90 percent of all wolf kills are elk; the other 10 percent consist of bison, deer, moose, and other small game.

RANCHERS' RESISTANCE TO WOLF REINTRODUCTION

Out in the U.S. West, ranchers control large tracts of land. Sometimes they own the land outright. Others lease the land from the U.S. government. The ranchers' concerns are basic. The wolf is a predatory animal that finds the easiest type of food source available. An animal that has been domesticated and no longer has natural predators is very easy prey.

From 1995 to 1998 nine head of cattle and 132 sheep were killed by wolves. The wolves that have killed livestock were mainly traveling from Canada to Yellowstone, across Montana. From 1987 to 1997 Defenders of Wildlife have paid $42,000 for 62 cattle and 141 sheep that have been lost to wolves. Many environmentalists feel that ranchers will kill off all of the introduced wolves. Only two wolves have died legally, while seven have died of unknown causes. The reintroduction of the wolf has had many problems, ranging from lawsuits to loss of livestock. The two lawsuits that have been filed contended it was unconstitutional to reintroduce the wolves into the park. The judge that was looking over the lawsuits said that the wolves needed to be returned to Canada, but Canada did not want them. Then the judge said that all the introduced wolves were to be sent to a zoo, but no zoo had room. Finally the judge said all of the introduced wolves needed to be destroyed, but the environmentalists protested. In the end nothing was done.

THE ROLE OF STATES

As of 2008 the population of the wolf has met the recovery goal of 10 breeding pairs. This means that the states are now trying to get the wolf off the endangered species list and under state control. Back around 1998 all of the states (Wyoming, Idaho, and Montana) started to make plans for how they were going to manage the wolf populations in their states. Each plan was then reviewed by wolf specialists and depredation specialists, to see what they thought of the plans. Each state had their plans finished by 2002. Wyoming was the first state to send its plan to be reviewed by the U.S. Congress. The other two states waited to see what would come of Wyoming's plan, before they sent theirs in. In 2003 the Wyoming Grey Wolf Management Plan was sent back to the state, saying that it would not work. Most people probably felt the plan did not go through because Wyoming was planning on managing the wolves. as if they were a predator species. This meant that the wolves could freely be hunted as long as they were off national forest or national park property and on private property. Many environmentalists did not want this since they felt this was the reason all the wolves had been lost in the first place. The Montana and Idaho plans were different than Wyoming's. They were planning on putting a trophy hunting season out on the wolf. As of right now, Wyoming has not made any changes to their plan, even though Congress wants them to change it to better manage the wolf population. Wyoming is going to take this matter to the courts.

POTENTIAL FOR FUTURE CONTROVERSY

This is a highly controversial topic at the local level that will continue to be debated in courts, legislatures, and federal agencies. As more species become endangered and protected, successful reintroduction programs around national parks and other federal lands will be initiated. The battleground for this controversy may move to legislatures. As federal and state wildlife and park agencies seek more funding for reintroduction programs others will seek to prevent this. Ranchers and others in surrounding communities will continue to resist the siting of dangerous animals in their midst. These very communities have often benefited from the presence of nearby national parks, as well as long-term grazing leases from the federal government at very favorable rates.

See also Endangered Species; Stock Grazing and the Environment

Web Resources

CNN.com. Wolves' Return to Yellowstone Sparks Controversy. Available at www.cnn.com/ EARTH/9711/12/yellowstone.wolves/. Accessed January 22, 2008.

Environmental Literacy Council. Wolves. Available at www.enviroliteracy.org/subcategory. php/292.html. Accessed January 22, 2008.

New West. The Greatest Hunting Controversy of Them All. Available at www.newwest.net/ index.php/topic/article/the_greatest_hunting_controversy_of_them_all/C146/L41/. Accessed January 22, 2008.

Further Reading: Mech, L. David, and Luigi Boitani. 2003. *Wolves: Behavior, Ecology and Conservation.* Chicago: University of Chicago Press.

WIND ENERGY SUPPLY

Wind power comes from turbines that generate power. These turbines can create noise that may disturb communities and wildlife. In cold climates the blades can throw ice and snow. Sometimes the variation in wind turbulence can cause traditional fan blades to come off. Some communities do not want them near homes or schools.

Proponents claim that wind power can be harnessed to be a nonpolluting, renewable source of energy to meet electric power needs around the world. Wind power is a form of renewable energy. As portions of the earth are heated by the sun, the heated air rises and air rushes to fill the low-pressure areas, creating wind. The wind is slowed as it brushes the ground so it may not feel windy at ground level. The power in the wind may be five times greater at the height of the blade tip on a large, modern wind turbine. Entire areas of the country may be very windy while other areas are relatively calm. The majority of people do not live in high-wind areas, although that could change with global warming and climate change.

WIND INTO ELECTRICITY

Wind power is converted to electricity as the wind moves the blades of a windmill or wind turbine. In a typical, modern, large-scale wind turbine, the wind is converted to rotational motion by the rotor, which is the three-bladed assembly at the front of the wind turbine. The rotor turns a shaft that transfers the motion into the large housing at the top of a wind turbine tower. The slowly rotating shaft enters a gearbox that greatly increases the rotational shaft speed. The output shaft is connected to a generator that converts the rotational movement into electricity.

APPLICATIONS

The wind resource in the United States is vast. Using today's technology, proponents claim there is theoretically enough wind power flowing across the United States to supply all of our electricity needs. Assuming adequate access to

---

One common controversy with earlier wind turbines was their noise. Noise issues are difficult because it is hard to measure noise in real-life conditions. The following are measurements in decibels of some common noises:

Source/Activity	Indicative noise level dB
Threshold of hearing	0
Rural nighttime background	20–40
Quiet bedroom	35
Wind farm at 350 m	35–45
Car at 40 mph at 100 m	55
Busy general office	60
Truck at 30 mph at 100 m	65
Pneumatic drill at 7 m	95
Jet aircraft at 250 m	105
Threshold of pain	140

While modern turbines are quieter, no research has been undertaken about the cumulative effects of constant noise from wind turbines.

---

the national grid, windy North Dakota alone could supply over 40 percent of the nation's electricity. Adequate winds for commercial power production are found at sites in all but four states. However, only a small portion of that potential is currently used. Less than 1 percent of U.S. electricity is currently supplied by wind power.

ENVIRONMENTAL IMPACTS

One of wind energy's important environmental benefits is its minimal impact on the environment. Electricity in the United States is mostly produced from coal and other fossil fuels (70%), nuclear energy (20%), and hydroelectric sources (dams), which have greater environmental impacts. There has been some controversy about the noise the wind turbines make when in operation.

Recently there has been major technological innovation in the design of multidirectional, conical turbines. Some claim to have designed them at a scale to fit on top of rooftops in urban areas. Cities such as Chicago have great interest in these "microturbines," but currently the skeptics are many. An emerging battleground is the government regulation of microturbines on urban rooftops. Given the density of the unit in urban areas it is unlikely that every building could harness the wind energy. The buildings will have to be able to take both the weight of the units and the stress of wind turbulence. Nonetheless, many engineers and builders think that lighter and stronger materials will be able to do so. There is promise of rapid technological advancement increasing the efficiency and safety of wind power.

POTENTIAL FOR FUTURE CONTROVERSY

As nonrenewable sources of energy dry up, other alternative sources must be found to replace them. Wind power is an attractive alternative. If developed on a large scale, community resistance could result from some of the noise concerns. Controversies may evolve as a national policy on alternative fuels gains a meaningful budget. How much should the government support wind power? Can a government use its power to take private property for wind turbines to generate power for community use? Would it matter if it were a private company, a utility, or a city that owned the wind turbines and/or land underneath it? These battleground questions and others have no easy answers yet, which is a sign of potential controversy.

New technological advancements in wind turbines and in other ways to effectively increase their efficiency may make them even more appealing. Some claim they can make them function on a building. Globally, wind markets are booming. The Global Wind Energy Council analyzed wind energy data from 70 nations. They report that in 2006 total installed wind energy capacity was 74,223 MW. In 2005 it was 59,090 MW. The wind energy market grew by 41 percent in 2006. Europe has the largest market share with 65 percent of the total. Germany and Spain are especially involved in wind energy, and alternative, renewable energy generally. Asia had 24 percent of new installations in 2006 according to the council report. Canada increased its wind power capacity enormously from 683 MW in 2005 to 1,459 MW in 2006. The United States has the highest new installed wind power capacity reporting 2,454 MW in 2006. Thirteen nations now generate at least 1,000 MW a year, and more are exploring the possibilities of wind power. Concern about energy independence as well as environmental impacts influences this battleground.

Given the recent rapid changes in technology of wind turbines and the robust increases in global wind energy markets, it is likely that any environmental controversies associated with them will increase. However, technology may be able to overcome concerns about noise and other risks, and decrease some of the current battlegrounds in this controversy.

See also Climate Change; Global Warming; Sustainability

Web Resources

Union of Concerned Scientists. Wind Power: Clean, Sustainable, and Affordable. Available at www.ucsusa.org/clean_energy/coalvswind/index.html. Accessed January 22, 2008.

U.S. Department of the Interior. Wind Energy Development Programmatic EIS Information Center. Available at windeis.anl.gov/. Accessed January 22, 2008.

Wind Power: Environmental and Safety Issues. Wind Energy Fact Sheet 4. Available at www.dti.gov.uk/files/file17777.pdf. Accessed January 22, 2008.

Further Reading: Gipe, Paul. 1995. *Wind Energy Comes of Age.* New York: John Wiley and Sons; Hills, Richard Leslie. 1994. *Power from Wind: A History of Windmill Technology.* Cambridge: Cambridge University Press; Pasqualetti, Martin J., Paul Gipe, and Robert Righter. 2002. *Wind Power in View: Energy Landscapes in a Crowded World.* St. Louis, MO: Elsevier; Patel, Mukund R. 1999. *Wind and Solar Power Systems.* New York: CRC Press.

APPENDIX A: ENVIRONMENTAL DATABASE PROGRAMS, APPLICATIONS, AND PORTAL WEB SITES

ABEL Model
http://www.epa.gov/compliance/civil/econmodels/index.html

By U.S. Environmental Protection Agency, Office of Enforcement and Compliance Assurance. The ABEL Model evaluates a corporation's or partnership's claim that it cannot afford compliance costs, cleanup costs, or civil penalties.

Arsenic Rule State Primacy Requirements (PDF)
http://www.epa.gov/safewater/arsenic/compliance.html#training

By U.S. Environmental Protection Agency, Office of Ground Water and Drinking Water. Slides and narrative from one of the presentations delivered in the EPA's 2002 Arsenic Rule Training Sessions.

Arsenic Rule State Primacy Requirements (PowerPoint)
http://www.epa.gov/safewater/arsenic/compliance.html#training

By U.S. Environmental Protection Agency, Office of Ground Water and Drinking Water. Copies of the slides from one of the presentations delivered in the EPA's 2002 Arsenic Rule Training Sessions.

BEN Model
http://www.epa.gov/compliance/civil/econmodels/index.html

By U.S. Environmental Protection Agency, Office of Enforcement and Compliance Assurance. The BEN Model calculates a violator's economic savings from delaying and/or avoiding pollution-control expenditures.

Dashboard of Sustainability
http://www.iisd.org/cgsdi/dashboard.asp

By Consultative Group on Sustainable Development Indicators. A software package that illustrates the complex relationships among economic, social, and environmental issues. The visual format is suitable for decision makers and others interested in sustainable development. This edition promotes the Millennium Development Goals (MDGs) indicators especially for developing countries. These indicators help define poverty reduction strategies and monitor the achievement of the MDGs.

Environmental Planning for Small Communities (TRILOGY)
http://www.epa.gov/grtlakes/seahome/trilogy.html

By U.S. Environmental Protection Agency and Purdue University. An introduction to a wide range of environmental issues and decisions that affect small to medium-sized communities. It offers communities the chance to judge their own needs and preferences and to make informed decisions on their own. Major sections cover: environmental laws and regulations, self-assessment planning and comparative risk analysis, financial tools and financial self-analysis, case studies, and contact and information directory. Each section includes interactive tools, such as a notebook tool to fill in and save survey forms and keep notes. The program combines and integrates numerous publications from federal agencies and states like Iowa, Minnesota, Montana, and Utah.

EnviroRisk
http://www.uic.edu/sph/cade/envirorisk/

By the School of Public Health at the University of Illinois at Chicago. A case-based, problem-solving program in environmental risk assessment and risk communication. You will play the role of a public health professional as you go through the course. This course will develop your ability to investigate an environmental health problem and serve as a resource and risk communicator in your community. Credit for taking this course is available through the Centers for Disease Control and Prevention.

Epi Info
http://www.cdc.gov/epiinfo/

By U.S. Centers for Disease Control (CDC). With Epi Info and a personal computer, epidemiologists and other public health and medical professionals can rapidly develop a questionnaire or form, customize the data entry process, and enter and analyze data. Epidemiological statistics, tables, graphs, and maps are produced with simple commands such as READ, FREQ, LIST, TABLES, GRAPH, and MAP. Epi Map displays geographic maps with data from Epi Info.

EpiCalc 2000
http://www.brixtonhealth.com/index.html

By Mark Myatt. A statistical calculator that works with pretabulated data. Some of the functions available in EpiCalc will be of general use, but they have been chosen to be of interest to persons working with data from a public health or epidemiological context.

HRS Quickscore
http://www.epa.gov/superfund/programs/npl_hrs/quickscore.htm

By U.S. Environmental Protection Agency, Superfund Program. Created to assist in scoring sites using the EPA's Hazard Ranking System (HRS). HRS Quickscore is an electronic set of HRS scoresheets that executes real-time site score calculations. It was designed to assist in developing a conceptual model for Superfund site assessments. This product is intended for use by those individuals who plan and implement preliminary assessments (PAs), site inspections (SIs), and other data-collection efforts according to the HRS rules, as well as those individuals that write and review HRS documentation records.

HRS Toolbox
http://www.epa.gov/superfund/sites/npl/hrsres/

By U.S. Environmental Protection Agency, Superfund Program. The HRS Toolbox page provides current guidance documents that may be used to determine if a site is a candidate for inclusion on the National Priorities List.

INDIPAY Model
http://www.epa.gov/compliance/civil/econmodels/index.html

By U.S. Environmental Protection Agency, Office of Enforcement and Compliance Assurance. The INDIPAY Model evaluates an individual taxpayer's claim that he or she cannot afford compliance costs, cleanup costs, or civil penalties.

MUNIPAY Model
http://www.epa.gov/compliance/civil/econmodels/index.html

By U.S. Environmental Protection Agency, Office of Enforcement and Compliance Assurance. The MUNIPAY Model evaluates a municipality's, town's, sewer authority's, or drinking water authority's claim that it cannot afford compliance costs, cleanup costs, or civil penalties.

National Emissions Inventory Input Format (NIF)
http://www.epa.gov/ttn/chief/nif/index.html

By U.S. Environmental Protection Agency. The NEI Input Format (NIF) is the format most widely used by state and local agencies to transfer data to the EPA's National Emission Inventory (NEI). The most current version of the NIF, Version 2.0, and all user documentation is posted here for users to download. In addition, there is a software program included here for download to help NIF users perform quality control (QC) checks on their files to ensure correct format specification.

National Management Measures to Control Nonpoint Source Pollution from
   Urban Areas
http://www.epa.gov/nps/urbanmm/

By U.S. Environmental Protection Agency, Office of Wetlands, Oceans, and Watersheds. This guidance document is designed to help citizens and municipalities in urban areas protect bodies of water from polluted runoff that can result from everyday activities. The guidance also offers help to states for implementing their nonpoint source control programs and to municipalities for implementing their Phase II Storm Water Permit Programs.

PROJECT Model
http://www.epa.gov/compliance/civil/econmodels/index.html

By U.S. Environmental Protection Agency, Office of Enforcement and Compliance Assurance. The PROJECT Model calculates the real cost to a defendant of a proposed supplemental environmental project.

Regulatory Economic Analyses Inventory
http://yosemite.epa.gov/ee/epa/eed.nsf/webpages/

By U.S. Environmental Protection Agency, National Center for Environmental Economics (NCEE). A database of a variety of regulatory economic analyses on most major U.S. regulations. Some of the recent ones are available for download.

RMP*Review
http://yosemite.epa.gov/oswer/ceppoweb.nsf/content/rmp_comp

By U.S. Environmental Protection Agency, Chemical Emergency Preparedness and Prevention Office (CEPPO). Designed for reviewing and analyzing risk management plans (RMPs) submitted under the Clean Air Act, Section 112(r). It has advanced query capabilities for users who want to analyze RMP data beyond the capabilities of RMP*Info (e.g., state implementing agencies, LEPCs, etc.)

STELLAR: Systematic Tracking of Elevated Lead Levels and Remediation
www.cdc.gov/nceh/lead/surv/stellar/stellar.htm

By U.S. Center for Disease Control (CDC). A software application provided to state and local childhood lead-poisoning prevention programs (CLPPPs) with a practical means of tracking medical and environmental activities in lead-poisoning cases. The intent of this application is to provide an electronic means of addressing the data that programs receive from labs, providers, clinics, and case management professionals.

Superfund Chemical Data Matrix (SCDM)
www.epa.gov/superfund/sites/npl/hrsres/tools/scdm.htm

By U.S. Environmental Protection Agency, Superfund Program. Contains factor values and benchmark values applied when evaluating potential National Priorities List (NPL) sites using the Hazard Ranking System (HRS). Factor values are used for determining the relative threat posed by a hazardous waste site and reflect hazardous substance characteristics, such as toxicity and persistence in the environment, substance mobility, and potential for bioaccumulation. Benchmarks are environment- or health-based substance concentration limits developed by or used in other EPA regulatory programs. SCDM contains HRS factor values and benchmark values for hazardous substances that are frequently found at sites evaluated using the HRS, as well as the physical, chemical, and radiological data used to calculate those values. PDF files.

Water Quality Standards and Criteria
www.epa.gov/grtlakes/seahome/wqs.html

By U.S. Environmental Protection Agency and Purdue University. Provides a framework for maintenance and improvement of water quality when adopted by states, U.S. territories, and Indian tribes. This software presents the three components of state and tribal water quality standards: water body uses (e.g., swimming, boating), water quality criteria or limits on chemical concentrations that may be present the water body, and antidegradation policy to protect existing water quality. It explains how Indian tribes can become involved and describes discretionary policies that affect water quality standards (WQS), economic considerations in the WQS program, the EPA's role and responsibility in the WQS process, and implementation of WQS through National Pollutant Discharge Elimination System (NPDES) permits. Relevant sections of the Clean Water Act are included, along with information on use attainability analysis, the submittal/approval process, implementation, and public involvement. Additionally, the user is provided with resources, contacts, a glossary of terms, and several examples of water quality standards.

Working with Communities for Environmental Health (Webcast on
   Demand)
www2a.cdc.gov/PHTNOnline/Registration/DetailPage

By U.S. Center for Disease Control (CDC). This program is an edited version of the Working with Communities for Environmental Health satellite broadcast and webcast, originally aired on September 12, 2002. Environmental conditions and their impact on health are a growing concern. It is challenging for communities and individuals to obtain reliable, accurate environmental health information. It is equally challenging for health and environmental professionals to work effectively with communities and individuals concerned about or facing environmental health issues. This program demonstrates a framework for working with communities and individuals to improve their capacity for making informed decisions that promote environmental health and quality of life. The program examines cultural, behavioral, environmental, and policy influences that impact community-based work throughout the process. Continuing education credits are available.

PORTAL WEBSITES

Air Quality in National Parks
http://www2.nature.nps.gov/air/Monitoring/network.cfm

Federal Facilities Environmental Stewardship and Compliance Assistance
   Center
http://www.fedcenter.gov/assistance/myfacility/

International Endangered Species List
http://news.nationalgeographic.com/news/

National Oceanic and Atmospheric Administration: Tides, historical
  nautical charts
http://www.noaa.gov/charts.html

U.S. Endangered Species List
http://www.fws.gov/endangered/wildlife.html

APPENDIX B:
INDEX CHEMICALS

Benzene and the polycyclic hydrocarbons are index chemicals—examples of the many chemicals present in combustion products. While the descriptions of the chemicals are very general, there is no doubt that chronic exposure can produce profound and long-lasting changes in biological function.

BENZENE

The greatest possibility for high-level exposures is in the workplace. Most people are exposed to benzene in tobacco smoke and automobile exhaust. Benzene has been found in at least 337 of 1177 National Priorities List (NPL) hazardous waste sites. Other environmental sources of benzene include gasoline (filling) stations, underground storage tanks that leak, wastewater from industries that use benzene, chemical spills, groundwater next to landfills containing benzene, and possibly some food products that contain benzene naturally. Brief Exposure at High Levels—Death may occur in humans and animals after brief oral or inhalation exposures to high levels of benzene; however, the main effects of these types of exposures are drowsiness, dizziness, and headaches. These symptoms disappear after exposure stops.

Long-Term Exposures at Various Levels—From overwhelming human evidence and supporting animal studies, the U.S. Department of Health and Human Services has determined that benzene is carcinogenic. Leukemia (cancer of the tissues that form the white blood cells) and subsequent death from cancer have occurred in some workers exposed to benzene for periods of fewer than 5 and up to 30 years. Long-term exposures to benzene may affect normal blood production, possibly resulting in severe anemia and internal bleeding. In addition, human and animal studies indicate that benzene is harmful to the immune system, increasing the chance for infections and perhaps lowering the body's defense against tumors. Exposure to benzene has also been linked with genetic changes in humans and animals. Animal studies indicate that benzene has adverse effects on unborn animals. These effects include low birth weight, delayed bone formation, and bone marrow damage. Some of these effects occur at benzene levels as low as 10 parts of benzene per million parts of air (ppm). Although benzene has been reported to have harmful effects on animal reproduction, the evidence for human reproductive effects, such as spontaneous abortion or miscarriage, is too limited to form a clear link with benzene. Benzene can be measured in the blood and the breath. The body changes benzene to phenol, which can be measured in the urine. Amounts of benzene (in blood) and phenol (in urine) cannot be used as yet to predict what degree of harmful health effects may occur. The Environmental Protection Agency (EPA) set the maximum permissible level in drinking water at 5 parts of benzene per billion parts of water (ppb). Because benzene can cause leukemia, the EPA established an ultimate goal of 0 ppb for benzene in drinking water and in ambient water such as rivers and lakes. The EPA realizes that this goal may be unattainable and has estimated how much benzene in ambient water would be associated with one additional cancer case for every 100,000 persons (6.6 ppb benzene), one case for every 1 million persons (0.66 ppb benzene), and one case for every 10 million persons (0.066 ppb benzene). The National Institute for Occupational Safety and Health (NIOSH) has recommended an occupational exposure limit in air of 0.1 part of benzene per million parts of air (ppm). The Occupational Safety and Health Administration's (OSHA) legally enforceable limit is an average of 1.0 ppm over the standard eight-hour workday, 40-hour workweek.

POLYCYCLIC AROMATIC HYDROCARBONS (PAHs)

PAHs are a group of chemicals that are formed during the incomplete burning of coal, oil and gas, garbage, or other organic substances. PAHs can be human-made or occur naturally. There is no known use for most of these chemicals except for research purposes. A few of the PAHs are used in medicines and to make dyes, plastics, and pesticides. They are found throughout the environment in the air, water, and soil. There are more than 100 different PAH compounds. Although the health effects of the individual PAHs vary, the following 15 PAHs are considered a group with similar toxicity: acenaphthene, acenaphthylene, anthracene, benz(a) anthracene, benzo(a)pyrene, benzo(b)fluoranthene, benzo (ghi)perylene, benzo (k)fluoranthene, chrysene, dibenz(a,h)anthracene, fluoranthene, fluorene, indeno(1,2,3-cd)pyrene, phenanthrene, pyrene. Several factors will determine whether harmful health effects will occur and what the type and severity of those health effects will be. These factors include the dose (how much), the duration (how long), the route by which you are exposed (breathing, eating, drinking, or skin contact), the other chemicals to which you are exposed, and your individual characteristics such as age, sex, nutritional status, family traits, lifestyle, and state of health. As pure chemicals, PAHs generally exist as colorless, white, or pale yellow-green solids. Most PAHs are found as mixtures of two or more PAHs. They can occur in the air either attached to dust particles, or in soil or sediment as solids. They can also be found in substances such as crude oil, coal, coal tar pitch, creosote, and road and roofing tar. Most PAHs do not dissolve easily in water, but some PAHs evaporate into the air. PAHs generally do not burn easily, and they will last in the environment for months to years. PAHs are attached to dust and other particles in the air and originate from vehicle exhausts, asphalt roads, coal, coal tar, wildfires, agricultural burning and hazardous waste sites. Background levels of PAHs in the air are reported to be 0.02–1.2 milligrams per cubic meter (mg/m^3) in rural areas and 0.15–19.3 mg/m^3 in urban areas. Exposure to PAHs can occur from soil near areas where coal, wood, gasoline, or other products have been burned or from the soil on or near hazardous waste sites, such as former manufactured-gas sites and wood-preserving facilities. PAHs have been found in some drinking water supplies in the United States. The background level of PAHs in drinking water ranges from 4 to 24 nanograms per liter (ng/L). For many people, the greatest exposure to PAHs occurs in the workplace. PAHs can enter the body through the lungs. They enter the body quickly and easily by all routes of exposure. The rate at which PAHs enter your body is increased when they are present in oily mixtures, and they tend to be stored in the kidneys, liver, and fat, with smaller amounts in the spleen, adrenal glands, and ovaries. Results from animal studies show that PAHs do not tend to be stored in the body for a long time and are excreted within a few days in the feces and urine. PAHs may be carcinogens. Several of the PAHs, including benz(a)anthracene, benzo(a)pyrene, benzo(b)fluoranthene, benzo(k)fluoranthene, chrysene, dibenz(a,h)anthracene, and indeno(1,2,3-cd)pyrene have caused tumors in laboratory animals when they ate them, when they were applied to their skin, and when they breathed them in the air for long periods of time. Reports in humans show that individuals exposed by breathing or skin contact for long periods of time to mixtures of other compounds and PAHs can also develop cancer. Mice fed high levels of benzo(a)pyrene during pregnancy had difficulty reproducing and so did their offspring. The offspring from pregnant mice fed benzo(a)pyrene also showed other harmful effects, such as birth defects and decreased body weight. Similar effects could occur in humans, but we have no information to show that these effects do occur. Studies in animals have also shown that PAHs can cause harmful effects on skin, body fluids, and the body's system for fighting disease after both short- and long-term exposure. These effects have not been reported in humans. PAHs are changed into chemicals that can attach to substances within the body. The presence of PAHs attached to these substances can then be measured in body tissues or blood after exposure to PAHs. However, this test is still being developed and it is not known yet how well it works. PAHs or their breakdown products can also be measured in urine. Although these tests can tell that you have been exposed to PAHs, it is not yet possible to use these tests to predict the severity of any health effects that might occur or to determine the extent of your exposure to the PAHs. These tests are not routinely available at a doctor's office because they require special equipment for sampling and detecting these chemicals.

For more information, see http://www.nutramed.com/environment/cars chemicals.htm.

APPENDIX C: GLOSSARY OF ENVIRONMENTAL TERMS

The following is a generally accepted set of terms used in describing environmental events.

Source: Agency for Toxic Substances and Disease Registry (ATSDR), U.S. Department of Health and Human Services.

Absorption: The process of taking in, as when a sponge takes up water. Chemicals can be absorbed into the bloodstream after breathing or swallowing. Chemicals can also be absorbed through the skin into the bloodstream and then transported to other organs. Not all of the chemical breathed, swallowed, or touched is always absorbed.

Acute: Occurring over a short time, usually a few minutes or hours. An acute exposure can result in short-term or long-term health effects. An acute effect happens within a short time after exposure.

Ambient: Surrounding. Ambient air usually means outdoor air (as opposed to indoor air).

Analyte: A chemical for which a sample (such as water, air, blood, urine, or other substance) is tested. For example, if the analyte is mercury, the laboratory test will determine the amount of mercury in the sample.

Aquifer: An underground source of water. This water may be contained in a layer of rock, sand, or gravel.

Background level: A typical level of a chemical in the environment. Background often refers to naturally occurring or uncontaminated levels. Background levels in one region of the state may be different than those in other areas.

Bedrock: The solid rock underneath surface soils.

Biological monitoring: Analyzing chemicals, hormone levels, or other substances in biological materials (blood, urine, breath, etc.) as a measure of chemical exposure, health status, etc. in humans or animals. A blood test for lead is an example of biological monitoring.

Body burden: The total amount of a chemical in the body. Some chemicals build up in the body because they are stored in body organs like fat or bone or are eliminated very slowly.

Case control study: A study in which people with a disease (cases) are compared to people without the disease (controls) to see if their past exposures to chemicals or other risk factors were different.

Central nervous system (CNS): The part of the nervous system that includes the brain and the spinal cord.

CERCLA: Comprehensive Environmental Response, Compensation and Liability Act. See "Superfund."

Chronic: Occurring over a long period of time, several weeks, months, or years.

Cohort study: A study in which a group of people with a past exposure to chemicals or other risk factors are followed over time and their disease experience compared to that of a group of people without the exposure.

Composite sample: A sample that is made by combining samples from two or more locations. The sample can be of water, soil, or another medium.

Concentration: The amount of one substance dissolved or contained in a given amount of another substance or medium.

Contaminant: Any substance that enters a system (the environment, human body, food, etc.) where it is not normally found. Contaminants are usually referred to in a negative sense and include substances that spoil food, pollute the environment, or cause other adverse effects.

Dermal: Referring to the skin. For example, dermal absorption means absorption through the skin.

Detection limit: The smallest amount of substance that a laboratory test can reliably measure in a sample of air, water, soil, or other medium.

Dose: The amount of substance to which a person is exposed.

Epidemiology: The study of the occurrence and causes of health effects in human populations. An epidemiological study often compares two groups of people who are alike except for one factor such as exposure to a chemical or the presence of a health effect. The investigators try to determine if the factor is associated with the health effect.

Exposure: Contact with a chemical by swallowing, by breathing, or by direct contact (such as through the skin or eyes). Exposure may be either short term (acute) or long term (chronic).

Exposure assessment: A process that estimates the amount of a chemical that enters or comes into contact with people or animals. An exposure assessment also describes how often and for how long an exposure occurred, and the nature and size of a population exposed to a chemical.

Feasibility study (FS): A study that compares different ways to clean up a contaminated site. The feasibility study recommends one or more actions to remediate the site. See "Remedial investigation."

Gradient: The change in a property over a certain distance. For example, lead can accumulate in surface soil near a road due to automobile exhaust. As you move away from the road, the amount of lead in the surface soil decreases. This change in the lead concentration with distance from the road is called a gradient.

Health assessment for contaminated sites: Determination of actual or possible health effects due to environmental contamination or exposure. It includes a health-based interpretation of all the information known about the situation. The information may come from site investigations (environmental sampling and studies), exposure assessments, risk assessments, biological monitoring, or health effects studies. The health assessment is used to advise people how to prevent or reduce their exposures, to determine remedial actions or the need for additional studies.

Health effects studies related to contaminants: Studies of the health of people who may have been exposed to contaminants. They include, but are not limited to, epidemiological studies, reviews of health status of people in exposure or disease registries, and doing medical tests.

Health registry: A record of people exposed to a specific substance (such as a heavy metal), or having a specific health condition (such as cancer or a communicable disease). New York State maintains several health registries.

Ingestion: Swallowing (such as eating or drinking). Chemicals in or on food, drink, utensils, cigarettes, hands, etc. can be ingested. After ingestion, chemicals may be absorbed into the blood and distributed throughout the body.

Inhalation: Breathing. People can take in chemicals by breathing contaminated air.

Interim Remedial Measure (IRM): An action taken at a contaminated site to reduce the chances of human or environmental exposure to site contaminants. Interim remedial measures are planned and carried out before comprehensive remedial studies. They can prevent additional damage during the study phase, but do not interfere in any way with the need to develop a complete remedial program.

Latency period: The period of time between exposure to something that causes a disease and the onset of the health effect. Cancer caused by chemical exposure may have a latency period of 5 to 40 years.

Leaching: As water moves through soils or landfills, chemicals in the soil may dissolve in the water thereby contaminating the groundwater. This is called leaching.

Maximum Contaminant Level (MCL): The highest (maximum) level of a contaminant allowed to go uncorrected by a public water system under federal or state regulations. Depending on the contaminant, allowable levels might be calculated as an average over time or might be based on individual test results. Corrective steps are implemented if the MCL is exceeded.

Media: Elements of a surrounding environment that can be sampled for contamination, usually soil, water, or air. Plants, as well as humans (when sampling blood, urine, etc.) and animals (such as sampling fish to update fish consumption advisories) can also be considered media. The singular of "media" is "medium."

Metabolism: All the chemical reactions that enable the body to work. For example, food is metabolized (chemically changed) to supply the body with energy. Chemicals can be metabolized by the body and made either more or less harmful.

Morbidity: Illness or disease. A morbidity rate for a certain illness is the number of people with that illness divided by the number of people in the population from which the illnesses were counted.

National Priorities List (NPL): A list maintained by the U.S. Environmental Protection Agency (EPA) of certain inactive hazardous waste sites. The list is produced and updated periodically by the EPA. See Superfund.

Odor threshold: The lowest concentration of a chemical that can be smelled. Different chemicals have different odor thresholds. Also, some people can smell a chemical at lower concentrations than others can.

Organic: Generally considered as originating from plants or animals, and made primarily of carbon and hydrogen. Scientists use the term organic to mean those chemical compounds that are based on carbon.

Permeability: The property of permitting liquids or gases to pass through. A highly permeable soil, such as sand, allows a liquid to pass through quickly. Clay has a low permeability.

Persistence: The quality of remaining for a long period of time (such as in the environment or the body). Persistent chemicals (such as DDT and PCBs) are not easily broken down.

Plume: An area of chemicals moving away from its source in a long band or column. A plume, for example, can be a column of smoke from a chimney or chemicals moving with groundwater.

Protocol: The detailed plan for conducting a scientific procedure. A protocol for measuring a chemical in soil, water, or air describes the way in which samples should be collected and analyzed.

Quality assurance and quality control (QA/QC): A system of procedures, checks, and audits to judge and control the quality of measurements and reduce the uncertainty of data. Some quality control procedures include having more than one person review the findings and analyzing a sample at different times or laboratories to see if the findings are similar.

Remedial investigation (RI): An in-depth study (including sampling of air, soil, water, and waste) of a contaminated site needing remediation to determine the nature and extent of contamination. The remedial investigation (RI) is usually combined with a feasibility study (FS).

Remediation: Correction or improvement of a problem, such as work that is done to clean up or stop the release of chemicals from a contaminated site. After investigation of a site, remedial work may include removing soil and/or drums, capping the site, or collecting and treating the contaminated fluids.

Risk: Risk is the possibility of injury, disease, or death. For example, for a person who has measles, the risk of death is one in one million.

Risk assessment: A process that estimates the likelihood that exposed people may have health effects.

Risk management: The process of deciding how and to what extent to reduce or eliminate risk factors by considering the risk assessment, engineering factors (Can procedures or equipment do the job, for how long, and how well?), social, economic, and political concerns.

Route of exposure: The way in which a person may contact a chemical substance. For example, drinking (ingestion) and bathing (skin contact) are two different routes of exposure to contaminants that may be found in water.

Safe: Strictly, free from harm or risk. Exposure to a chemical usually has some risk associated with it, although the risk may be very small. However, many people use the word safe to mean something that has a very low risk or one that is acceptable to them.

Site inspection: A Department of Health visit to a site to evaluate the likelihood of human exposure to toxic chemicals and to do an exposure assessment.

Solubility: The largest amount of a substance that can be dissolved in a given amount of a liquid, usually water. For a highly water-soluble compound, such as table salt, a lot can dissolve in water. Motor oil is only slightly soluble in water.

Superfund (federal and state): The federal and state programs to investigate and clean up inactive hazardous waste sites.

Target organ: An organ (such as the liver or kidney) that is specifically affected by a toxic chemical.

Volatile: Evaporating readily at normal temperatures and pressures. The air concentration of a highly volatile chemical can increase quickly in a closed room.

Volatile organic compound (VOC): An organic chemical that evaporates readily. Petroleum products such as kerosene, gasoline, and mineral spirits contain VOCs. Chlorinated solvents such as those used by dry cleaners or contained in paint strippers are also VOCs.

GENERAL BIBLIOGRAPHY

Agrawal, Clark C. 2001. *Communities and the Environment: Ethnicity, Gender, and the State in Community-Based Conservation.* Piscataway, NJ: Rutgers University Press.

Boyce, James K., and Barry G. Shelley, 2003. *Natural Assets: Democratizing Ownership of Nature.* Washington, DC: Island Press.

Bullard, Robert. 1993. *Confronting Environmental Racism: Voices from the Grassroots.* Boston: South End Press.

———. 2000. *Dumping in Dixie: Race, Class, and Environmental Quality.* Boulder, CO: Westview Press.

Cole, Luke W., and Sheila R. Foster. 2001. *From the Ground Up: Environmental Racism and the Rise of the Environmental Justice Movement.* New York: New York University Press.

Collin, Robert W. 2006. *The U.S. Environmental Protection Agency: Cleaning Up America's Act.* Westport, CT: Greenwood Press.

Common, Michael S. 1995. *Sustainability and Policy: Limits to Economics.* Cambridge: Cambridge University Press.

Cox, John D. 2005. *Climate Crash: Discovering Rapid Climate Change and What It Means to Our Future.* New York: National Academies Press.

Daniels, Ronald J., Donald F. Kettl, and Howard Kunreuther, eds. 2006. *On Risk and Disaster: Lessons from Hurricane Katrina.* Philadelphia: University of Pennsylvania Press.

Dean, Bartholomew, Jerome M. Levi, and Winona LaDuke. 2003. *At the Risk of Being Heard: Identity, Indigenous Rights and Postcolonial States.* Ann Arbor: University of Michigan Press.

Degregori, Thomas R. 2004. *Origins of the Organic Agriculture Debate.* Ames, IA: Blackwell Publishing.

Depoe, Stephen P., and John W. Delicath. 2004. *Communication and Public Participation in Environmental Decision Making.* Albany, NY: SUNY Press.

Dernbach, John C. 2000. *Stumbling toward Sustainability.* Washington, DC: Environmental Law Institute.

Doob, Leonard William. 1995. *Sustainers and Sustainability: Attitudes, Attributes, and Actions for Survival.* Westport, CT: Praeger.

Durning, Alan. *Poverty and the Environment.* Portland, OR: Worldwatch Institute.

Easton, Thomas A., and Theodore D. Goldfarb, eds. 2003. *Taking Sides: Clashing Views on Controversial Environmental Issues.* Dishkin, CT: McGraw-Hill.

Elliott Johansen, Bruce. 2003. *The Dirty Dozen: Toxic Chemicals and the Earth's Future.* Westport, CT: Praeger.

Environmental Protection Agency. 1997. *Ecological Risk Assessment Guidance for Superfund: Process for Designing and Conducting Ecological Risk Assessments.* Washington, DC: U.S. Environmental Protection Agency.

Forman, Richard T. T. 2002. *Road Ecology: Science and Solutions.* Washington, DC: Island Press.

Frank, Lawrence D., Peter O. Engelke, and Thomas L. Schmid. 2003. *Health and Community Design: The Impact of the Built Environment on Physical Activity.* Washington, DC: Island Press.

Freeman, A. Myrick III. 2003. *The Measurements of Environmental and Resource Values: Theory and Methods.* Washington, DC: Resources for the Future.

Freyfogle, Eric T. 2003. *The Land We Share: Private Property and the Common Good.* Washington, DC: Island Press/Shearwater Books.

Friedman, Frank B. 2003. *Practical Guide to Environmental Management.* Washington, DC: Environmental Law Institute.

Garwin, Richard L., and Georges Charpak. 2002. *Megawatts and Megatons: The Future of Nuclear Power and Nuclear Weapons.* Chicago: University of Chicago Press.

Geisler, Charles, and Gail Daneker, eds. 2000. *Property and Values: Alternatives to Public and Private Ownership.* Washington, DC: Island Press.

Haar, Charles M., and Jerold S. Kayden. 1989. *Zoning and the American Dream: Promises to Keep.* Chicago: American Planning Association.

Harkin, Michael Eugene, and David Rich Lewis. 2007. *Native Americans and the Environment: Perspectives on the Ecological Indian.* Lincoln: University of Nebraska Press.

Harrad, Stuart. 2001. *Persistent Organic Pollutants: Environmental Behaviour and Pathways of Human Exposure.* New York: Springer.

Honey, Martha. 1999. *Ecotourism and Sustainable Development: Who Owns Paradise?* Washington, DC: Island Press.

Houck, Oliver A. 1999. *The Clean Water Act TMDL Program: Law, Policy, and Implementation.* Washington, DC: Environmental Law Institute.

Institute of Medicine. 2000. *Clearing the Air: Asthma and Indoor Air Exposures.* Washington, DC: National Academies Press.

Kan, Sergei. 2006. *New Perspectives on Native North America: Cultures, Histories, and Representations.* Lincoln: University of Nebraska Press.

Kreske, Diori L. 1999. *Environmental Impact Statements: A Practical Guide for Agencies, Citizens, and Consultants.* New York: John Wiley and Sons.

Lawrence, David Peter. 2003. *Environmental Impact Assessment: Practical Solutions to Recurrent Problems.* Hoboken, NJ: John Wiley and Sons.

Maser, Chris. 1999. *Ecological Diversity in Sustainable Development: The Vital and Forgotten Dimension.* Washington, DC: Lewis Publishers.

Mazaika, Rosemary, Robert T. Lackey, and Stephen L. Friant, eds. 1995. *Ecological Risk Assessment: Use, Abuse, and Alternatives.* Amherst, MA: Amherst Scientific Publishers.

Mech, L. David, and Luigi Boitani. 2003. *Wolves: Behavior, Ecology and Conservation.* Chicago: University of Chicago Press.

Meltz, Robert, Dwight H. Merriam, and Richard M. Frank. 1999. *The Takings Issue: Constitutional Limits on Land-Use Control and Environmental Regulation.* Washington, DC: Island Press.

Moore, Colleen F. 2003. *Silent Scourge: Children, Pollution, and Why Scientists Disagree.* New York: Oxford University Press.

Moussiopoulos, Nicolas, ed. 2003. *Air Quality in Cities.* New York: Springer.

Nash, Linda Lorraine. 2007. *Inescapable Ecologies: A History of Environment, Disease, and Knowledge.* Berkeley: University of California Press.

National Academies Press. 2003. *Cumulative Environmental Effects of Oil and Gas Activities on Alaska's North Slope.* Washington, DC: National Research Council.

National Environmental Education Advisory Council. 2005. *Setting the Standard, Measuring the Results, Celebrating Successes: A Report to Congress on Environmental Education in the US.* Washington, DC: U.S. Environmental Protection Agency.

National Environmental Justice Advisory Council (NEJAC). 1996. *The Model Plan for Public Participation.* Washington, DC: Environmental Protection Agency.

Noss, Reed F., 1997. *The Science of Conservation Planning: Habitat Conservation under the Endangered Species Act.* Washington, DC: Island Press.

O'Riordan, Timothy. 2006. *Interpreting the Precautionary Principle.* London: James & James/Earthscan.

Pelling, Mark. 2003. *The Vulnerability of Cities: Natural Disasters and Social Resilience.* London: Earthscan.

Pellow, David, and Robert J. Brulle. 2005. *Power, Justice, and the Environment.* Cambridge, MA: MIT Press.

Platt, Rutherford H. 2004. *Land Use and Society: Geography, Law, and Public Policy.* Washington, DC: Island Press.

Randolph, John. 2004. *Environmental Land Use Planning and Management.* Washington, DC: Island Press.

Riddell, Robert. 2004. *Sustainable Urban Planning: Tipping the Balance.* Oxford: Blackwell Publishing.

Selin, Helaine. 2003. *Nature across Cultures: Views of Nature and the Environment in Non-Western Cultures.* New York: Springer.

Shiva, Vandana. 2002. *Water Wars: Privatization, Pollution, and Profit.* Boston: South End Press.

Simioni, Daniela. 2004. *Air Pollution and Citizen Awareness.* New York: United Nations Publications.

Spellerberg, Ian F. 2005. *Monitoring Ecological Change.* Cambridge: Cambridge University Press.

Stamati, P. Nicolopoulou. 2004. *Cancer as an Environmental Disease*. New York: Springer Publishing.

Steingaber, Sandra. 1997. *Living Downstream: A Scientist's Personal Investigation of Cancer and the Environment*. New York: Random House.

Thomas, June Manning, and Marsha Ritzsdorf. 1997. *Urban Planning and the African American Community: Planning in the Shadows*. Thousand Oaks, CA: Sage Publications.

Tickner, Joel A. 2002. *Precaution, Environmental Science, and Preventive Public Policy*. Washington, DC: Island Press.

United Nations Environment Programme. 1997. *Global Environment Outlook*. Oxford: Oxford University Press.

Victor, David G. 2001. *The Collapse of the Kyoto Protocol and the Struggle to Slow Global Warming*. Princeton, NJ: Princeton University Press.

Wackernagel, Mathis, and William Rees. 1995. *Our Ecological Footprint: Reducing Human Impact on the Earth*. Gabriola Island, Canada: New Society Publishers.

Wargo, John. 1998. *Our Children's Toxic Legacy: How Science and Law Fail to Protect Us from Pesticides*. Yale University Press.

Yost, Nichlas C. 1989. *NEPA Deskbook*. Washington, DC: Environmental Law Institute.

ABOUT THE AUTHOR AND CONTRIBUTORS

Steven Bonnoris Fellow, Public Interest Law Research Institute, Hastings College of Law. Over the past several years he has managed a series of reports jointly produced by the Public Interest Law Research Center and the American Bar Association, including "Environmental Justice for All: A Fifty State Survey." He recently authored *Environmental Enforcement in the Fifty States: The Promise and Perils of Supplemental Environmental Projects.* He is a graduate of Harvard College and Harvard Law School.

Robert W. Collin is the Senior Research Scholar at the Center for Sustainable Communities at Willamette University. He has been a Professor of Law, Urban Planning and Social Work; teaching at the University of Auckland, New Zealand; the University of Virginia Department of Urban and Environmental Planning, the University of Oregon Environmental Studies, Cleveland State University Department of Social Work, and Jackson State University Department of Urban and Regional Planning. He has served as an advisor to state and federal agencies. He has published many articles. His most recent book is *The US Environmental Protection Agency: Cleaning Up America's Act.*

Robin Morris Collin Professor of Law, Willamette School of Law. Professor Morris Collin has numerous publications in the area of sustainability. She was the first law professor to teach sustainability in the United States in the early 1990s. She also teaches cultural property law and has published in this area. She has litigated court cases, served as an advisor to federal and state agencies, and provided legislative testimony on many environmental and cultural issues. She is working with the Oregon State Bar Association to find ways to integrate sustainability into legal practice.

Cathy Koehn is a retired teacher and a past director of the Northwest Cougar Action Trust, a nonprofit that actively worked to ban hounding and baiting of cougars and bears. She has a bachelor's degree in social sciences and a master's degree in science education. She has taught classes at Western Washington University and worked for the Forest Conservation Council, which produced forest ecosystem maps for the Audubon Society. She has been actively involved in saving Oregon's wildlife since the mid 1980s and continues to make regular public comment to the state's Wildlife Commission and at legislative hearings on wildlife and predator issues.

Monica Patel is a 2006 law school graduate of Lewis and Clark Law School. She is currently working on issues of Environmental Justice in Michigan.

INDEX